THE THREAT TO EARTH

How Climate Change Has Become The Life And Death For Humanity

Christopher Uchenwa

10-10-10 Publishing

THE THREAT TO EARTH:
How Climate Change Has Become The Life And Death For Humanity.
www.gewo-intl.org/the-threat-to-earth/

Paperback ISBN: 978-1-7389222-4-6
Hardcover ISBN: 978-1-7389222-1-5
eBook ISBN: 978-1-7389222-2-2
Audio ISBN: 978-1-7389222-3-9

Publisher
10-10-10 Publishing
Markham, ON Canada

Printed in Canada and the United States of America

To my dear brother Pius Uchenwa, who passed away on March 1, 2023, and whose life was a beacon of kindness, courage, and love. Your unwavering support and unbreakable spirit inspired me to pursue my dreams, and I dedicate this book to you as a tribute to the legacy you left behind. Although you are no longer with us, your memory will continue to shine brightly through the pages of this book and in the hearts of those who loved you. Rest in peace, dear brother.

TABLE OF CONTENTS

Acknowledgements

I am grateful to God Almighty, the Creator of the universe. So many years ago, you put this burden in my heart. You refused to give up on me even when I did everything to run away from pursuing this purpose. You gave me the wisdom and the power to accomplish this dream. Thank you very much.

I appreciate the scientists, innovators, and activists whose tireless efforts have propelled us forward in scientific knowledge and sustainable practices. Their work has paved the way for us to create a better future for our planet.

To Evelyn Mae and her wonderful team, as well as Naval Kumar, my book architect, I extend my heartfelt appreciation for your contributions in making this dream a reality.

I would also like to express my deepest thanks to Mr. Raymond Aaron, *New York Times* top 10 best-selling author, my personal mentor and coach, and his team for their invaluable assistance in bringing this book to life.

To all those who have supported me throughout this journey, I am grateful beyond words. While there are many more people I would like to thank, I must stop here out of respect for time, space, and modesty.

Foreword

As the threat of climate change looms ever larger, it's more important than ever to take action to protect your planet. That's where Christopher Uchenwa comes in. As an award-winning Nigerian-Canadian author, environmental activist, and expert in sustainability accounting, Christopher has dedicated his life to creating a safer world by protecting the environment from the dangers posed by global warming.

With a background in accounting and finance, Christopher brings a unique perspective to the fight against climate change. He is a member of both the Chartered Professional Accountant (CPA) of Canada and the Association of Chartered Certified Accountants of United Kingdom, and his expertise in sustainability accounting gives him unparalleled authority in the subject matter of his book. Christopher's childhood experiences of the devastating effects of pollution and deforestation in Nigeria inspired his passion for environmental activism.

In *The Threat to Earth,* Christopher explores the urgent need for action to save the planet from the devastating effects of global warming. He provides a comprehensive overview of the causes and effects of climate change, from rising temperatures and melting polar ice caps to extreme weather events and the decline in biodiversity. He also examines the impact of climate change on your health, including increased air pollution, disease outbreaks, and displacement and submerging of cities due to extreme weather events and rising sea levels.

Furthermore, Christopher looks at the economic consequences of climate change, such as the loss of coastal infrastructure and properties, decline in agricultural production, and energy costs. He also examines the threat of poor nations in the global fight against climate change, and the importance of international cooperation and policy action.

Overall, *The Threat to Earth* is a thought-provoking and informative book that highlights the grave threat of climate change to your planet and those who live here with you. It urges you to take immediate action to address the crisis, and emphasizes that it is only through individual and collective efforts that you can hope to create a sustainable future. I'm honored to have been Christopher's personal mentor and coach, and I'm proud to recommend this book to you, who cares about the future of your planet.

Raymond Aaron
***New York Times* Bestselling Author**

Chapter 1

Introduction

Encounter with the Creator of Planet Earth – The Light Being

I was alone in the heart of the desert, shivering as the cold night breeze sliced through me. The moon above was like a sickle for harvesting wheat, casting a faint glow over the barren landscape. Three stars sparkled in the sky, one shining brighter than the others. I wished for their warmth, but it was futile. I was ill-prepared for the brutal cold. I heard the sound of the cold night breeze, whistling through my ears like it had a message for me. As the cold breeze sifted through my skin, I heard the continuous whisper like the whistling pine permeating my every being. The more I huddled on the ground, the more I realized how vulnerable I was. The desert offered no shelter, no escape. I felt a sense of dread wash over me, as if I was facing something far worse than the bitter cold. It was indeed cold out here! Then I began to shiver as the fear of death crept into my mind and my body began to give in to the cold.

In my desperation to stay warm and stay alive, I looked around me for safety, but all I could see was a desert of sand dunes, empty lands, and dry rocky mountains ahead of me. My hope was growing thin as I shivered and waited for the cold hand of death to take my soul from this cold ravaged body. There was silence all around me and there was darkness too. I was all alone by myself in this desert. As I lie on the ground, my eyes continued to look at the moon and her three little glimmering stars. Those were the only source of light that existed in this godforsaken desert. Something inside of me was looking up to the moon for help, but I knew it was a wishful thought. No one and nothing was coming to save me in this desert. I was alone and I was going to die from the brutal hands of the cold that kept squeezing the life gradually out of me.

I closed my eyes and opened them again as I gazed up to the moon. I noticed that something strange happened. The brightest of the three stars began to move in a circular form around the moon. I thought perhaps my mind was playing tricks on me. I closed my eyes again. This time I kept them closed for 10 seconds and hoped my dizziness would go away. When I opened my eyes again and looked at the moon and the stars, the brightest of the three stars was now moving faster around the moon. At first, I thought it was my imagination playing tricks on me. But as the star grew closer, I saw it was no illusion. It was hurtling towards me like a shooting star, its fiery tail streaming behind it. I was too weak to run, too frozen to move. I lay there, helpless, as the star blazed a path towards me.

Like a movie, I watch the star falling at a very fast speed, and I saw splinters of lights dissipating and scattering in the air as the star cut through the thick darkness like a hot knife cutting through margarine without any hindrance. It was heading directly towards me like a missile positioned to hit its target. Now I knew it was too late for me to do anything. I was too weak and cold to move or even try to run for safety. I just lay there watching the falling star and waiting for it to disintegrate me into oblivion. At least it gave me comfort to know that my suffering would end fast and in the blink of an eye.

As I lay there on the cold ground and awaited my death from the shooting star, a lot of thoughts began to quickly slip into my mind. I remembered the research I once did a long time ago out of curiosity, to know about the origin of shooting stars. In ancient times, people observed shooting stars and comets and often created mythological explanations to account for their appearance. One classical narration of how shooting stars or comets happen comes from Greek mythology.

According to the ancient Greeks, shooting stars were the result of the goddess Eris, the goddess of strife, tossing her golden apple into the sky. The apple, which was inscribed with the words "for the fairest," caused a quarrel among the goddesses Hera, Athena, and Aphrodite, who all claimed it for themselves. To settle the dispute, the goddesses turned to the mortal Paris to decide who should have the apple. Paris, who was promised the love of the most beautiful woman in the world in exchange for his decision, ultimately chose Aphrodite as the fairest. As a result, Hera and Athena were angry and hurled their weapons at the sky, creating the streaks of light that we now call shooting stars.

Comets, on the other hand, were believed to be a sign of impending doom in ancient times. The Greeks believed that comets were the fiery chariots of the gods, sent to warn mortals of impending disasters such as war or plague. They were also thought to be the embodiment of the god Hermes, the messenger of the gods, who would ride his flaming chariot across the sky to deliver important messages to the people of Earth.

While we now know that shooting stars and comets have scientific explanations that don't involve gods and goddesses, these ancient myths are a testament to our human fascination with the mysteries of the universe and our desire to understand the world around us.

I needed a more scientific explanation, so I dug further and discovered that shooting stars, also known as meteors, are caused by small pieces of space debris, typically no larger than a grain of sand or a pebble, entering the Earth's atmosphere at high speed. As they travel through the atmosphere, they create friction with the air molecules, which causes the debris to heat up and produce a streak of light in the sky. This light is what we see as a shooting star.

Comets, on the other hand, are much larger objects than the debris that causes shooting stars. Comets are made up of a mixture of ice, dust, and small rocky particles and are often described as "dirty snowballs." As a comet approaches the sun, the heat causes the ice in the comet to vaporize, producing a glowing cloud of gas and dust around the nucleus of the comet. This glowing cloud, known as the coma, can be visible from Earth as a fuzzy patch in the sky. As the comet gets closer to the sun, some of the dust and gas in the coma is blown back by the solar wind, creating a tail that can extend for millions of kilometers. The tail is what we see as a comet in the night sky.

Both shooting stars and comets are fascinating phenomena that capture our imagination and curiosity. While we now understand the scientific explanations for these events, they continue to inspire wonder and awe in people of all ages. I found that there are larger objects than the shooting stars and comets that also fall to the earth. They are called meteorites.

A meteorite is a solid object that originates in space and reaches the surface of the Earth. Meteorites are typically fragments of asteroids or other objects in the solar system that have been dislodged from their parent bodies by collisions or other processes.

When a meteoroid, which is a small, rocky or metallic body, enters the Earth's atmosphere, it is traveling at very high speed, typically tens of thousands of miles per hour. As it encounters the Earth's atmosphere, it compresses the air in front of it, causing the air to heat up and the meteoroid to become incandescent. This incandescent object is what we see as a shooting star. Most meteoroids are small and burn up completely in the atmosphere, producing no lasting effects on the ground.

However, if the meteoroid is large enough to survive its passage through the atmosphere, it will impact the Earth's surface and become a meteorite. Meteorites can be made of different materials, including rock, metal, or a combination of both. They can vary in size from tiny pebbles to massive boulders weighing several tons.

Meteorites are important to scientists because they provide valuable information about the formation and evolution of the solar system. By analyzing their composition, age, and other properties, scientists can learn about the conditions that existed when the meteorites were formed, as well as the processes that have affected them since their formation. For example, some meteorites contain tiny grains of stardust that predate the solar system, providing clues about the early history of our galaxy.

As I lay there, still watching the shooting star that was heading towards me, I was convinced it was a meteorite sent by the gods to ease my pain and suffering. Death at that point could be a glorious passage for me. The cold was mean and brutal on my skin. I remembered that was the third day of my water-only fast. I am not a religious person; rather, I would say I align more towards spirituality. Oftentimes, once every quarter, I usually embark on a 3-day water-only fast. My reason for such exercise was to subject my physical body through a natural process of cleansing and rejuvenation. I observed that fasting to me is like rebooting my computer to refresh it and clear every background running program that is slowing and making my computer function slowly and clumsily. Fasting clears my brain and mind and makes me think more clearly and sharply. In addition, it leaves me healthier as it forces my body's mechanism to eat up and flush out waste, dead cells, excess sugar, and cholesterol in my system that could make me sick, cancerous, or clumsy. On the spiritual side, I found that my intuitive perception is heightened, my gut decision making around my life and business is enhanced, and every form of depression and negative energy is neutralized and flushed away.

My First Encounter with "The Light Being"

It was day 3 of my water-only fasting and, by my calendar, it was January 3, 2017. I had been sleeping all night and everything happening to me was a lucid dream. It was so real as my consciousness was involved in this lucid dream. I felt as though that night was going to be my last night on Earth as I watched the falling shooting star heading towards me to destroy me and snatch the life out of me. It was close to me now, too bright for my eyes to keep gazing, so I closed my eyes to receive the impact of the collision. It happened so quickly, and when I opened my eyes again, the shooting star had transformed into a Light Being and was standing some few meters away from me. Its legs were not touching the ground. It was as if it was floating in the air underneath its feet. I noticed that the being was so bright and had the shape of a translucent human. Its skin was as colorless as that of water. I could see through its body form, which was shaped like that of a human, with the help of the bright light emitting around its body form. I couldn't look straight into it as the light was too bright for my eyes to behold, so I looked away.

Something else happened to me. I noticed that suddenly I was feeling warm and the cold inside of me had disappeared. It was a good feeling that was replaced with fear. Then it was like I heard a voice that seemed to be referring to me. The voice was so sweet, soft, and healing to my soul. "Do not be afraid, the keeper of truth," was the voice I heard. It was so clear. I asked, "Please, who are you?" The voice coming from the Light Being responded, "I am the wisdom that created everything and the Earth you live in. I have an assignment for you."

"What assignment?" I asked. The sweet-sounding voice responded, "You are the keeper of truth. It is in your life's path to deliver this message to the inhabitants of the Earth. Go tell humankind to *save the Earth now or risk losing humanity!*"

At this point, I was getting more confused. I managed to ask again, "Save the Earth from what?" The Light Being reached out and touched me on the shoulder while I was still coiled up on the ground out of fear. It said, "Come with me; I want to show you something." The moment It touched me, I noticed some changes happened to me. I could now look at it directly without the bright shiny light affecting my eyes. It looked like a translucent being of light in the form of a human. Its face was like that of the most beautiful woman I have ever seen in my life, while its body was like that of a masculine male warrior. It looked so beautiful and glorious, and I felt my inner spirit being drawn to it like the two opposite poles of a magnet trying to attract each other. Its presence around me was so calming, peaceful, and glorious.

The moment it said to come with it, I found myself and it on the top of a very tall mountain. Then it began to speak again. "I am the wisdom that created all things and your planet Earth. I am going to explain to you how your Earth was formed. In the beginning, when I created the Earth, the Earth was without form and covered by water. There was no land. Then I separated the water from the land by pushing the waters to the north pole of the Earth and froze a good portion of the waters into ice and glaciers to make room for land, upon which I was going to create humans, animals and other creatures. I left 70% of the Earth to be waters, and the remaining 30% to be land." In my head, I was beginning to say, "Wow, wow, wow. This story is beginning to look like those of the biblical book of Genesis. I summoned the courage and asked the Light Being, "Is this the same story in the book of Genesis of the Bible, and it said, "Yes. However, I am going to give you more details so that you will understand your assignment very well."

The Light Being went on speaking in its usual soft and calming manner. "I created the sun to provide heat energy to the Earth and all creatures in it. Humans and other creatures living on planet Earth would depend on the sun for their source of life and sustenance forever." I managed to ask, "How do you mean?" It went on to say, "The energy of the sun would provide warmth to the Earth as well as food to every living thing on the Earth. I created the green plants and vegetation to trap the energy coming from the sun and convert that energy into food to sustain all the living creatures on Earth." At that point, I knew it was talking about the process of photosynthesis. Back in the days in my agriculture and biology class, I learned so much about photosynthesis and the Earth's ecosystem.

Photosynthesis is the process by which plants, algae, and some bacteria convert light energy from the sun, into chemical energy that can be stored as glucose or other sugars. The process can be divided into two stages:

Light-dependent reactions:
In this stage, the light energy is absorbed by the chlorophyll pigment in the plant's leaves and is used to generate ATP (adenosine triphosphate) and NADPH (nicotinamide adenine dinucleotide phosphate). These energy-rich molecules are used in the second stage of photosynthesis.

Light-independent reactions (Calvin cycle):
In this stage, the ATP and NADPH generated in the first stage are used to convert carbon dioxide (CO_2) into glucose or other sugars. This process is called the Calvin cycle, which takes place in the chloroplasts of the plant cell.

The Calvin cycle involves a series of chemical reactions that result in the conversion of CO_2 to glucose. The cycle begins with the fixation of CO_2, which is the process of incorporating carbon from CO_2 into an organic molecule. The fixed carbon is then converted into a series of intermediate molecules, which are eventually used to produce glucose or other sugars. The Calvin cycle also requires energy in the form of ATP, which is generated in the light-dependent reactions.

Overall, photosynthesis is a vital process for life on Earth, as it provides the basis for most of the food chains and ecosystems. It also releases oxygen into the atmosphere, which is essential for the survival of many organisms, including humans.

"Does it now make sense to say that the sun is the source of life to all living things on Earth?" I asked the Light Being. It said, "Yes, the sun can be considered the source of life on Earth. Without the energy provided by the sun, life as we know it would not exist. The sun provides the light and heat that allows plants to perform photosynthesis, which is the process by which they convert carbon dioxide and water into oxygen and glucose. Oxygen is essential for the survival of most animals, including humans, and glucose is the primary source of energy for all living organisms.

In addition to providing energy for photosynthesis, the sun also plays a crucial role in regulating the Earth's climate and weather patterns. The sun's radiation heats the Earth's atmosphere and oceans, creating winds and ocean currents that distribute heat and moisture around the planet. This, in turn, creates the conditions for diverse ecosystems and weather patterns.

Overall, the sun is an essential component of the Earth's ecosystem, and without its energy and radiation, life on Earth would not be possible."

The Light Being went on to say, "Now, let me tell you how I positioned the sun and the Earth to sustain life. I created the atmosphere over the Earth's surface and made it in such a way that it would trap the carbon dioxide released by humans and other living organisms during their respiratory process. You could consider the atmosphere as a blanket or sheet over the Earth that is made up of majorly carbon dioxide and other gases. This carbon dioxide gas and other gases in the atmosphere is meant to trap the sun's heat energy that comes into the Earth, so that the Earth would be kept warm and be able to sustain life. Without this atmospheric layer, the Earth's surface would be so cold, and no living things would be able to survive. In order to create a balance in the atmospheric gases, I made the plants and green vegetation to absorb some of the carbon dioxide for the purpose of their own respiratory process, and in return release oxygen back into the atmosphere for humans and other creatures to utilize for their own respiratory process. This is a balanced and harmonious ecosystem that I put in place to ensure the sustenance of life forms on planet Earth.

The Light Being turned and looked at me. It asked, "Is this making sense to you now?" I quickly said yes and went on to tell him how amazed I was at the wisdom behind the creation of the Earth. I noticed that it smiled and glowed brighter as it said to me, "I am the infinite intelligence behind all creations. When I was done creating the Earth, humans, and all creatures on the Earth, I blessed humans and gave them five major mandates. He paused and looked at me to be sure I was ready to receive the next thing he was going to say. Then he continued. "The five mandates I gave to humans after creation of the Earth are to be fruitful, multiply, subdue the Earth, dominate the Earth, and replenish the Earth. The Light Being paused again and looked at me, and I looked at him blankly. It said, "I know you really do not understand the details of these mandates. Don't worry; I am going to break it down for you."

The Light Being went on to say, "These five mandates that I gave to humans are already recorded in the book of Genesis. I also refer to them as the Creation Mandate, and they are meant to provide insight into the relationship between me, the infinite wisdom behind creation, humanity, and the natural world, the Earth. They emphasize the importance of procreation, population growth, responsible management of the natural world, wise leadership, and environmental stewardship. They provide guidance for ethical behavior and responsible decision-making regarding the Earth. In clearer detail, here are what these five mandates stand for:

Be Fruitful: The first mandate, to "be fruitful," means that I, the Creator, intended for humans to grow and thrive both physically and spiritually. This mandate is a call to procreation, both biologically and spiritually, and it emphasizes the importance of passing on faith and values to future generations. Humans have been very procreative since creation; however, humans are yet to attain their true spiritual superhuman nature. The Light Being went on to reveal much deeper wisdom to me as to why humans are yet to attain their true spiritual superhuman nature. For the purpose of staying focused on the objective of this book, I have decided to delve more into the other details relating to the spiritual superhuman nature, in my upcoming book titled, *The Missing Secrets.*

Multiply: "The second mandate, to 'multiply,' is closely related to the first," the Light Being said. This mandate emphasizes the importance of population growth and the continuation of the human species. It also emphasizes the importance of cultivating relationships and building community. Humans have been able to accomplish this mandate. The population of the Earth has grown over 8 billion since creation, and I am impressed. However, humans are gradually losing the ability to relate with one another as they move further away from the world of reality. At this point, I knew exactly what the Light Being was referring to. The world is moving from natural reality to virtual reality, and the metaverse technology is quickly taking us there. Humans are gradually losing personal relationship and communal living with one another, and this will weaken the love bond amongst humanity.

Subdue the Earth: The Light Being continued talking in its usual beautiful calm voice. The third mandate, to "subdue the earth," means that I intended for humans to take an active role in managing and developing the natural world. This mandate is a call to stewardship, which includes responsible management and preservation of the Earth's resources. This means that humans are called to manage and care for the Earth and its resources, including animals, plants, and the environment. This involves using the earth's resources for human benefit, but also being responsible stewards of the natural world, avoiding exploitation or harm. So far, humans have taken this as a license to exploit and dominate the natural world, leading to environmental degradation, destruction, and threat to Earth.

Dominate the Earth: While I was wondering if I was really dreaming or going insane, I heard that calm sweet voice again as the Light Being continued. The fourth mandate, to "dominate the Earth," means that humans have been given a position of authority over the natural world. This mandate is a call to leadership and responsibility, and it requires humans to exercise wisdom and discernment in the use of power.

On the one hand, humans have made significant progress in harnessing the Earth's resources and building civilizations that have improved the quality of life for many people. You have developed technologies that allow you to travel, communicate, and produce goods and services more efficiently than ever before. You have built infrastructure, created agricultural systems, and developed medicines that have helped you to thrive and survive.

However, on the other hand, there are many ways in which humans have failed to meet the call to responsible stewardship of the Earth. Your exploitation of the Earth's resources has led to significant environmental degradation, including pollution, deforestation, soil erosion, and climate change. You have also contributed and are still contributing to the extinction of many species of living things, and caused harm to ecosystems that are critical to sustaining life on Earth.

Additionally, there are social and ethical issues related to the way humans have "dominated" the Earth. The benefits of your progress have been distributed unequally, with some people and communities benefiting more than others.

The activities of humans in dominating the Earth irrationally is not without its consequences. In the exercise of power, humans have often exploited people and resources for personal gain or profit. This has led to social and economic injustice, including the unequal distribution of resources, exploitation of vulnerable populations, and violations of human rights.

Humans have often made decisions that prioritize short-term gain over long-term consequences. This has led to actions that harm the environment, society, and future generations.

In some cases, humans have used their power to benefit themselves or a small group, rather than serving the common good. This has led to inequality and social unrest, undermining the well-being of society.

In summary, humans have often failed to fulfill the dominion mandate by exercising poor leadership, responsibility, and wisdom and discernment in the use of power. These failures have resulted in significant harm to the environment, society, and future generations, and highlight the need for a renewed commitment to responsible stewardship and care for the Earth and its inhabitants.

"Replenish the Earth," the Light Being said in a little louder tone. This is the fifth mandate that I gave to humans. It means you are to restore and renew the natural world. This mandate is a call to environmental stewardship, which includes reforestation, soil conservation, and other practices that promote the health and vitality of the Earth.

Humans are to ensure that the Earth and its resources are sustained and renewed. It involves the responsible use of natural resources and the development of sustainable practices to ensure that future generations can also benefit from the Earth's abundance.

In a deeper sense, this applies to the restoration of damaged ecosystems and the protection of endangered species. It is a call to repair the damage that has been done to the Earth and to ensure that it remains a healthy and vibrant home for all its inhabitants.

"Have you not noticed that the Earth needs an urgent replenishment?" the Light Being said while looking at me directly in the eyes. At this time, I was even more confused on what this whole thing was leading to. I knew deep down inside of me that there's a serious problem. The Light Being continued, "Climate change, pollution, and the loss of biodiversity are just a few of the issues that threaten the sustainability of the Earth's resources. The mandate to replenish the Earth calls on humans to take responsibility for their impact on the environment and to develop sustainable practices that ensure the health and vitality of the Earth and its inhabitants for generations to come.

Then the Light Being paused and looked at me, and I then heard a voice in my head saying, "Now ask all the questions running through your mind." I quickly asked, "Why are you telling me all this?" Then it said, "Because it is on your life's path to warn humanity. Take this message to them and tell them, *'Save the Earth now, or risk losing humanity.'"* Then I asked, again, "How do you mean lose humanity?" The Light Being looked at me as if it was disappointed in that question. Then it said, "Have you not understood all I have been saying all this time?

Okay, let me break it down. The blanket covering over the Earth, which you call the Earth's atmosphere, has been distorted as a result of the activities of humans. There is a lot more carbon dioxide and other gases in the atmosphere now, which is causing the Earth to trap more heat energy that penetrated the Earth from the sun. The more heat energy the Earth's atmosphere holds, the hotter the Earth will become. If this continues, it will overheat the Earth and cause all the water—which I trapped as glaciers and ice sheets in the poles in Antarctica and Greenland, and as smaller glaciers and ice caps in high altitude regions around the world, such as the Andes Mountains and the Himalayas—to begin to melt. When this happens, the bodies of water will begin to rise and will eventually cause the Earth to submerge and be swallowed up by water.

"What is the proof for all these things you're saying," I asked. The Light Being answered me even before I finished asking the questions. "Humans are intelligent beings. Go see the evidence for yourself in the work and investigations that NASA has been doing on climate change." I said, "NASA?" It said, "Yes, the National Aeronautics and Space Administration has been monitoring these changes and the deterioration of the Earth's climate condition. I had a bit of relief now that I was certain that the Light Being had been talking specifically about climate change and the impending danger that awaits the world. I quickly said, "There are a lot of advocates for climate change, so why do you need me to jump in on the train? Moreso, I am just a chartered professional accountant with a specialty in rendering financial system solutions as a consultant. There are people out there more qualified that are already pursuing this course of advocating for the danger of climate change. The Light Being then touched my shoulder and said to me, "It's on your life's path to pursue this course as well. Everyone on this path has a unique message. Your message is to warn humanity of the impending danger. As a matter of fact, in the next 100 years to come, some major cities could be submerged and disappear from the face of the Earth."

"What cities?" I asked.

"Major coastal cities and islands across the continents shall be affected. Go and do your research and you shall come up with the answers. Major cities and islands in North America, Europe, Asia and Africa shall disappear as a result of climate change in the next 100 years. Some of this might happen sooner than expected."

I woke up abruptly from my sleep. I was sweating even though the room was cold. At least I was glad it was all a dream. Somewhere in my mind, I was disturbed and worried at the same time. It was a lucid dream and it seemed to me like I was in a cinema watching a movie. I was conscious of every detail of the dream, and I remembered every detail. From my experience, I know some of my dreams do come true. I remember, in 2002, when I saw a man that looked like my dad waving me goodbye in my dreams. He was fair in complexion and dressed in chieftaincy clothing with a staff in his hand. I watched him pass me on the same road, and when I turned to look at him, he waved goodbye to me and kept on walking. I stopped and kept looking at him. When he got some good distance away, he fell to the ground and his chieftaincy staff fell out of his hands. The ground opened up and swallowed him. I screamed, "Nooooooooo!" I woke up frightened. That same day, I got a call that my dad had *kicked the bucket.* He had been ill for a while and did not make it.

This was the same frightened feeling I had when I woke up from sleep after my first encounter with the Light Being. I quickly rushed to my laptop and opened the NASA website to read about some of the work they had been doing on climate change. The first sentence I saw on the NASA website was as follows:

"How do we know climate change is real?
There is unequivocal evidence that Earth is warming at an unprecedented rate.
Human activity is the principal cause."

This statement caught my attention and interest the most. It sounded exactly like what the Light Being had said in my dream in my first encounter. I took my time going through the evidence and practical data being provided on the NASA website. The more I read, the more scared I became, as the evidence corroborated everything the Light Being had told me in my dream.

Still curious for more information, I quickly opened the software Bible on my phone. The Bible needed some time to update. I rarely read the Bible ever since I became spiritual. To me, most of the stories in it don't add up, and they conflict with what I consider spiritual truth. That is a story I wish to tell in my other upcoming book, *The Missing Secret.* I began to read from Genesis Chapter 1. The more I read, the more the creation story in the Bible began to make sense. I saw that in the beginning, the Earth was 100% water, until God separated the water from the land. I believe the separation happened simply by capturing a good portion of the water in the North Pole region of the Earth, which scientists called the Artic region. Another baffling verse in the Bible that captured my attention was Genesis Chapter 1:28, which talks about what God did after creating humankind. It says:

Then God blessed them, and God said to them, "Be fruitful and multiply; fill the earth and subdue it; have dominion over the fish of the sea, over the birds of the air, and over every living thing that moves on the earth."

This verse seems to be talking about the five mandates the Light Being explained vividly to me in my dream. There is more to this verse than I can imagine. This was where we were handed over the stewardship of the Earth by the landlord; and from all indication, we have not been doing a good job of taking care of the Earth.

MY Second Encounter with the Light Being

I woke up with a start, gasping for air as if I had just come up for air after being submerged in water for a long time. My body was drenched in sweat, and my heart was pounding so hard that I could feel it in my chest. I sat up in bed, looking around my darkened room as I tried to calm myself down. My heart was still racing like I had just run a marathon. I lay there, trying to catch my breath, wondering what had just happened. I looked around my room, trying to get my bearings. Everything looked familiar, but I couldn't shake off the feeling of terror that had consumed me just moments before.

As I caught my breath, the memory of my dream came flooding back to me. I had met the God of creation, the same Light Being that seemed to radiate power and majesty. But the meeting had quickly turned into a nightmare as a terrible flood had swept through the land, threatening to drown me and everything I held dear.

I shuddered as I remembered the feeling of being pulled under the water, the sound of rushing water filling my ears. It had felt so real, as if I had been transported to another world. I looked at the clock on my bedside table and it said 4:30 a.m., Wednesday, January 3, 2021. It quickly occurred to me that it was exactly four years to the day since I'd had the first encounter with the strange godlike being in my dream.

The first encounter happened on Tuesday, January 3, 2017. That first encounter had been different, but it had also been a warning. The Light Being had warned me of the impending doom of climate change, urging me to take action to warn humanity to save the Earth now or risk losing planet Earth and humanity. After my first encounter, I discarded it as a dream or rather a figment of my imagination. However, to play on the safe side of caution, I went ahead and opened a Facebook page somewhere around March 2017, and I named it "Global Environmental Watch Organization (GEWO)." For more information, visit: https://www.facebook.com/GEWOInternational.

I kicked off the parent chapter in Canada and a local chapter of the organization in Nigeria, and I enlisted some members to send in some environmental pollution evidence affecting people's health. The results that came in were alarming. A major city in Nigeria, Port Harcourt to be precise, was experiencing air pollution from oil and gas flaring. When people would wipe their nostrils or body with a white handkerchief or tissue, black sooty dirt messed up the white handkerchief or tissue. Doctor Kenny Adinnu, whom I enlisted to pioneer GEWO-Nigeria, confirmed to me that people were beginning to fall ill from this pollution. We kicked off some advocacy and awareness campaigns to prevent the worst from happening, but it did not go far as I and my team were self-sponsoring the whole initiatives. Financing the initiative became my biggest constraint. At least I was happy I tried to heed the warning of the Light Being, but I had found it difficult to convince others to take the threat seriously.

My second encounter with the godlike Light Being in my dream was more or less like a meeting for an update. I was on a vacation in a peaceful coastal town enjoying a sunny day, with people playing on the beach, swimming in the ocean, and walking along the promenade. Suddenly, a massive earthquake hit the ocean floor, causing a huge displacement of water that created a massive tsunami.

The people on the beach were initially confused and startled by the sudden surge of water. They had only a few moments to react as the tsunami approached rapidly. Chaos and panic ensued as people started running in different directions, trying to find higher ground. I immediately joined the race for dear life.

As the tsunami slammed into the shore, it unleashed a wall of water that swept away everything in its path. Buildings, boats, and vehicles were crushed and tossed around like toys, while people were dragged underwater and struggled to stay afloat. The powerful waves continued to batter the town, causing widespread devastation.

At this time, I was struggling to stay afloat as the heavy water current tossed me all over like a weightless feather. I was drowning and about to give up the fight when I felt a hand grab me by the shoulder, and I found myself high in the air above the water and massive flood. I looked around and there it was—the same godlike Light Being I had seen in my dream four years ago. It was glowing with light all around its colorless body, which looked like that of water. It took me to a very high mountain where I felt safe again. The first word it spoke to me was, "What have you done since the last time we met?" The sound of its voice was still as beautiful and calming as the last time. I was proud to tell him how I had started a Facebook group to create awareness on climate change. It looked at me with disappointment on its face, and it said to me, "You've not yet started. You do not understand the assignment before you."

The Light Being started walking towards the right side of the mountain and asked me to follow it. I observed that its legs were not touching the ground. It was actually "float-walking," for lack of a better word. It then turned to me and asked again, "Did you remember what happened in 1991 when you were leading a Young Foresters Advocacy group? How old were you then?"

I said I was 15 years old, and it asked me to tell about it. I began to tell it how my secondary school, in a sub-Saharan Africa, appointed me to attend a World Bank assisted afforestation program to combat desert encroachment and protect the environment. "What happened after the program?" it asked. I continued on how I formed the Young Foresters Club in my school, and we embarked on a tree planting project across the state then. He asked me again, "Why did you stop planting trees? Why did you stop advocating for a cleaner environment through tree planting?"

I responded that when I got into university, I did not deem it worthwhile to continue the initiative as it was capital intensive. Then it said to me, "Can you see that your life's path found you early enough, but you left it along the way? If you keep justifying your excuses, you will never get anything done in life. Excuses are the reason the grave is full of many unaccomplished dreams. It's time to take responsibility for the future of your planet and to act now to save the Earth. The clock is ticking, and the stakes couldn't be higher. Will you rise to the challenge, or will you face the unthinkable consequences of your inaction? The choice is yours. Go tell the world to act now to save the Earth, or risk losing humanity forever!" Those were the last words I heard, and when I turned around, it was gone.

I looked down from the mountain top and I saw the survivors struggling to cope with the aftermath of the tsunami. They began to realize that the disaster had even more far-reaching consequences. The massive displacement of water caused by the tsunami had a ripple effect that extended to the poles of the Earth.

The massive waves triggered by the earthquake also had an impact on the glaciers in the Arctic and Antarctic. The sudden increase in water temperature caused by the tsunami weakened the glaciers, causing them to melt at an alarming rate.

As a result, massive amounts of water were released into the ocean, which caused sea levels to rise rapidly. The coastal towns that had been hit by the initial tsunami were now facing an even greater threat, as the rising sea levels caused more flooding and destruction. I noticed everything was beginning to submerge under water, including the tall mountain I was on.

I woke up from this startling dream sweating. Now, as I sat in bed, still trembling from the after-effects of my dream, I realized that I needed to double up on my efforts. I couldn't shake the feeling that the dream had been a warning again, just like the one I had received four years ago. I made a mental note to start working on my environmental advocacy projects again, hoping that it wasn't too late to make a difference. I went on to register the Facebook Global Environmental Watch Organization as a nonprofit in Canada, with the plan of setting up chapters in every country in the world, to help create a more impactful awareness and solution.

I knew that I couldn't ignore the messages that I was receiving, no matter how strange or otherworldly they seemed. I vowed to do everything in my power to prevent the catastrophe that my dreams were warning me about.

I got out of bed and walked around my room, trying to shake off the feeling of unease that had settled over me. I splashed some cold water on my face and took a deep breath, trying to calm my racing heart.

As I got dressed and prepared to face the day, I couldn't help but feel a sense of dread. The dream had left me feeling unsettled and anxious, and I couldn't shake off the feeling that something bad was going to happen. I tried to tell myself that it was just my imagination, but the feeling wouldn't go away.

Throughout the day, I found myself constantly looking over my shoulder, wondering if I was being watched. I jumped at every unexpected sound, feeling like I was still being chased by the flood from my dream that tried to drown me.

It wasn't until the next night, when I had a peaceful sleep, that I finally felt like I was back to my normal self. But the memory of that terrifying dream stayed with me for a long time, a reminder of how powerful the mind can be, even when we're asleep.

My Third Encounter with the Light Being

On January 3, 2023, I was on the third night of my three-day, water-only fast when I had a dream. I was sitting in a Starbucks coffee shop in downtown Toronto, sipping my hot coffee and working on my laptop, when a young man with a cute dress style and Afro natural dreadlocks hair walked up to me and asked politely, "Can I join you here?" I said yes, and he sat down beside me.

After a few moments, he started talking about how people have politicized climate change. He then asked me if I believed in climate change, to which I replied that I did. I explained that I had registered a nonprofit organization to advocate for the dangers of climate change and how the whole world needs to come together to save our planet. The young man seemed interested and asked me to tell him more about it.

He asked me if I was reaching out to more people and if I thought I was doing well in my advocacy. I told him that I thought I wasn't taking it too seriously, and that I had other activities occupying my time. He then asked if I had a website, to which I replied that I had been planning to build one.

He asked me what had been holding me back, and I told him that it seemed like a huge project that would require significant resources to pull off. Additionally, a lot of people don't believe in climate change, and I didn't want to commit my scarce resources to it and end up getting frustrated by the politics around it. What if no one believes me? What if a lot of people criticize me? What if I fail in convincing the world about this impending danger?

The young man smiled and asked me if I believed climate change is real. When I said yes, he replied that that was all I needed to surmount every other thing holding me back. He suggested that I put all my research findings into a book and use the book to raise funds to drive this project, set up a nonprofit website, and get people who also believe in climate change to support my project. While you may encounter obstacles and criticism, in the end, you will overcome them and succeed.

He urged me to give it my best shot and ignore those who don't believe, focusing instead on those who do, so that together, we can save the Earth from the impending danger. As he spoke, his suggestions struck me as bright ideas. I was just about to ask him who he was and what he did for a living when he said, "You were born for this purpose and for this mission. I have made you a climate god to save the world from the impending danger. Go tell the world to save the Earth now, or risk losing humanity forever!"

He placed his hand on my shoulder and said, "I ordered another cup of coffee for you. Go get it when the order is ready." Then he walked away the same way he came in. I was still dazed and just realized that this was the same Light Being that had appeared to me twice in the past and now in human form. When I came out of the shock, he was gone. Then I heard my name from the Starbucks partner, "Chris, your coffee is ready!" I actually did hear my name as I woke up from sleep. It was my wife calling me to come downstairs for breakfast.

I went about my normal business as usual and promised myself to take my climate change advocacy a step further. However, the scary thought of encountering the so-called creator of the Earth, three times in my dream, just wouldn't let go. I kept thinking, what if my dreams were real and I had the responsibility to warn the world of the impending danger, and I did nothing about it? Would I ever be able to forgive myself?

It was March 1, 2023. I received a call from my brother's wife that my elder brother had passed away. The news shattered my spirit, and I was broken. Just the day before, I had spoken to the doctor of the private hospital where he was admitted. She was explaining my brother's test results to me and the treatments he had been undergoing. The doctor recommended a transfer to a specialist, a neurosurgeon, the next day as his test result was able to diagnose brain cancer, which he had been battling with for some time now. Unfortunately, my brother did not make it to the next morning when the transfer was going to take place. By 3:00 a.m., in the early hours of the morning, he passed away.

I cried every night in my silent moments by myself as I mourned my brother. It was like my world was standing still. A lot of thoughts began to go through my mind. My elder brother died at the young age of 56, leaving a wife and 4 beautiful children behind. I remembered some of his dreams he had shared with me that never came to life. He had wanted to own a jazz club and a community library as these were his two favorite hobbies. But death took him too soon to realize his dreams. Each night, I cried in my study for two weeks while I reminisced about the sweet moments we had shared together while growing up. With warm tears dripping down my face, I wrote him a poem for his funeral, to capture our sweet moments. His death woke me up from my slumber regarding my dreams. His death taught me that whatever I have to do in life, I should do it now. Today, you're here. Tomorrow, you may be no more. Live each day like it's your last. I now remembered that I had a mission to fulfill, a life path I was created to follow. I jumped back into my nonprofit advocacy on climate change, and I wrote this book, *The Threat to Earth*, as a courtesy of the wakeup call from the loss of my brother. That is why I have dedicated the book to him, Pius Uchenwa, my beloved brother.

Adieu My Beloved Brother Pius – March 17, 2023

Life was tough! You fought bravely, but in the end, death won the battle in the physical realm. However, you have transitioned into victory and gone to be with your Maker, waiting for your return. You and I went through this journey together, and I wish you could have held on a little longer...

I remember how you started off so well at an early age. You became an excellent air traffic controller. When I was 15 years old, you took me to live with you in Kano State. I was still in my early teenage years. You gave me a good start in life, a proper foundation that launched me into life and made me the person I am today.

gewo-intl.org/the-threat-to-earth/

It was from you that I learned to love reading books, because you loved reading novels. It was from you that I learned to appreciate jazz and classical music, because the sweet melodies always woke me up from sleep every morning.

You introduced me to great jazz players like Earl Klugh. He was your favorite jazz artist and became mine as well. As I pen these memories in my diary, I am listening to your best tracks of Earl Klugh. I know you're nodding your head right now as "Calypso Getaway" is playing. I know you're probably making those dance moves right now as the "Time After Time" track kicks in. Then, "Wishful Thinking," "If You're Still in Love with Me," "Kari," "Night Drive," "All the Time," "Cast Your Fate to the Wind," "Hotel California," "This Time," "Midnight in San Juan," "Another Time, Another Place," "Tropical Legs," "All I have to do is dream,"….

Each time I listen to these Earl Klugh tracks, sweet memories of you flood my mind. I remember how I used to smile each time your friends called you "Okey Jazzman." You inspired jazz music so much that your love for jazz became infectious to people around you. I must confess, I contracted the jazz virus from you.

You were indeed a deep thinker, an intelligent and gentle soul. You were a good man and would never hurt a fly.

As I listen to "Calypso Getaway," one of your best Earl Klugh's tracks, a sweet smile engulfs my face as warm tears run down my cheeks. Sweet memories of you miming this track came flooding into my mind. "Okey Jazzman" kept resonating in my head. Oh, Pius, how I miss you. You were such a good soul and gone too soon.

I remember how you introduced me to novels: "Gone with the Wind" by Margaret Mitchell, "The Third Eye" by Lobsang Rampa, and all the collections of James Hadley Chase. I still remember your top best were:

- *You Find Him, I'll Fix Him*
- *The Vulture Is a Patient Bird*
- *No Orchids for Miss Blandish*
- *An Ace up My Sleeve*
- *Come Easy, Go Easy*

- *You Never Know with Women*
- *The Whiff of Money*
- *You're Dead without Money*
- *The Way the Cookie Crumbles*
- *The Guilty Are Afraid*
- *There's Always a Price Tag*
- *The Sucker Punch*
- *I'll Bury My Dead*

I want to go back and read every one of these novels again. That was how we used to have our own fun then. We competed with James Hadley Chase novels. It was so much fun that if you owned only one James Hadley Chase novel, you could read all the different series by exchanging with other people. It was an "own one, own all" system.

Pius, you laid a foundation that made me who I am today. If only I could turn back the hands of time, I would bring you back to life, because I believe you were yet to deliver to the world the wealth of wisdom buried inside of you.

You lived in your world in your mind. Maybe you were introverted. A lot of people did not understand you. They saw your gentility, and many took advantage of you. Why? Because you were a good, gentle soul.

Oscar Olise saw a brother in you. I love the friendship that existed between both of you. It broke him when I relayed the news of your demise. He lost yet another good brother.
Adieu, my beloved brother Pius!

I believe you shall return to complete the purpose you left behind. Wherever you are now, may you find peace with God, your source and maker. Adieu, my beloved brother Pius!

I fought to save you. I fought with the might and determination of the warrior king of our ancestors, but death had an upper hand. When I saw you in my dream, whole and hearty, I did not know you came to say goodbye. I thought it was victory.

gewo-intl.org/the-threat-to-earth/

I am convinced the heritage you left behind will attain victory eventually, in the battle of life you started.

Adieu, my beloved brother Pius!
I love you so much, and I miss you too.
You will forever be remembered.
Adieu, my beloved brother Pius!

In the Beginning – School of Thought on How the Earth Came About

The Earth

The Earth is a planet that orbits the sun and is the third planet from the sun. It is the only known planet to support life and is therefore sometimes referred to as the "blue planet," due to the dominant color of its surface, which is 70% covered by water, with the remaining 30% being landmass.

The Earth has a diameter of approximately 12,742 kilometers and a circumference of 40,075 kilometers. It has a surface area of approximately 510 million square kilometers and is divided into continents and oceans. The Earth has a magnetic field that protects it from the harmful solar and cosmic radiation, and a large, rocky moon that orbits it.

The Earth's atmosphere is composed of 78% nitrogen, 21% oxygen, and 1% other gases, including carbon dioxide and water vapor. The atmosphere helps regulate the temperature of the Earth and protects it from the impact of meteoroids.

The Earth's rotation and its position relative to the sun causes the planet to have seasons, and its axis tilt causes the changes in the length of daylight throughout the year. The planet takes 365.25 days to complete one orbit around the sun, resulting in one year calendar. Its rotation takes 24 hours to complete one full turn, resulting in one day and night.

The Earth has a diverse landscape, including mountains, valleys, deserts, forests, and oceans. It is home to a wide variety of plants, animals, and microorganisms that form complex ecosystems.

The planet also has a hydrosphere, consisting of water in its various forms, including oceans, lakes, rivers, and groundwater. This water is essential for life on Earth and helps regulate the planet's temperature.

The Earth is able to sustain life due to a combination of several factors, including its location in the solar system, its atmosphere, and its water resources.

Location in the solar system: The Earth is located at the right distance from the sun, where it receives just the right amount of sunlight to support life. If the Earth were closer to the sun, it would be too hot for most life forms to survive, and if it were farther away, it would be too cold.

Atmosphere: The Earth's atmosphere acts like a protective blanket, shielding the planet from harmful solar radiation and allowing for the retention of heat and the creation of a suitable climate for life. The atmosphere is also composed of various gases, including nitrogen and oxygen, which are necessary for the survival of many living organisms.

Water resources: The Earth is the only known planet with large amounts of liquid water, which is essential for life. Water is used by living organisms for various functions, such as regulating body temperature, transporting nutrients, and serving as a habitat for aquatic life.

Geological stability: The Earth's solid crust and molten core provide a stable platform for life to thrive. The dynamic geological processes, such as plate tectonics and volcanic activity, help to regulate the planet's climate and create a diverse range of habitats for living organisms.

In summary, the unique combination of the Earth's location in the solar system, its atmosphere, water resources, and geological stability, make it a suitable environment for life to exist and thrive.

How did the Earth come about?

There are different schools of thought as to the origin of the Earth; and by choice, feel free to stick to whichever one gives you comfort to believe. The focus of this book is more about the threat to Earth than its origin. That notwithstanding, it will do us better to understand the organized structure and ecosystem of the Earth in order to appreciate the threats under which it is currently facing.

There is the scientific and religious school of thought on how the Earth came about.

The exact process of how the Earth was created is still a subject of scientific investigation and debate. However, the prevailing scientific theory is that the Earth was formed about 4.5 billion years ago from the dust and gas left over after the formation of the sun. This material eventually coalesced to form the Earth and the rest of the solar system.

According to this theory, the early Earth was a hot, molten mass that gradually cooled and solidified. As it cooled, heavy elements such as iron and nickel sank to the center of the Earth to form the core, while lighter elements rose to the surface to form the mantle and crust. Over time, geological processes such as volcanic eruptions, earthquakes, and the movements of tectonic plates created the Earth's landforms and shaped its surface. Another theory, known as the "big impact" theory, or the "big bang" theory, suggests that the Earth was formed from the collision of two smaller bodies in the early solar system. This collision would have created a massive amount of heat and pressure, which would have caused the Earth to melt and then re-form.

My concern with the scientific school of thought about how the Earth was formed is that science is yet to prove the reason for the order that resulted from the "big bang." Scientifically, after an explosion, the particles and matter that made up the explosion would assume a state of disorder, characterized by a high degree of randomness and disorganization. In the immediate aftermath of an explosion, the energy released by the explosion would cause the particles and matter to move in all directions at high speeds, creating a chaotic and disordered state.

We can recall the nuclear explosions in Hiroshima and Nagasaki that took place in August of 1945 during the second world war. On August 6th, the United States dropped a nuclear bomb on the city of Hiroshima, causing widespread destruction and killing an estimated 140,000 people. Three days later, on August 9th, the US dropped a second nuclear bomb on the city of Nagasaki, killing an estimated 70,000 people.

These bombings marked the first and only time in history that nuclear weapons have been used in armed conflict, and they had a profound impact on the world and on the course of human history. The devastating power of the bombs and the catastrophic loss of life they caused helped to bring an end to World War II and sparked a global effort to prevent the spread and use of nuclear weapons.

Till date, some people are still suffering the effects of the radioactive fallout from the nuclear explosions of Hiroshima and Nagasaki. While the immediate aftermath of the explosions was catastrophic, with many people being killed or injured by the blast and subsequent fires, the long-term effects of radiation exposure have also had a significant impact on the health of survivors and their descendants.

Studies have shown that the survivors of the bombings, known as Hibakusha, have experienced a higher rate of certain types of cancer and other illnesses as a result of their exposure to radiation. Additionally, some of their children and grandchildren have also experienced health problems that are thought to be related to the radiation exposure of their parents or grandparents.

The Japanese government has established programs to provide medical and financial support to the survivors of the bombing, but the long-term effects of radiation exposure continue to be a source of concern for many people. Additionally, the cities of Hiroshima and Nagasaki continue to monitor the levels of radioactive isotopes in the environment, and work to minimize the risks to public health posed by residual radioactive contamination.

Overall, while much progress has been made in the decades since the bombings of Hiroshima and Nagasaki, the radioactive fallout from the explosions continues to have a lasting impact on the health of those who were exposed to the radiation, and it serves as a powerful reminder of the devastating consequences of nuclear weapons.

The disorderliness that results after an explosion is evidence enough to disprove the scientific "big bang" theory as the origin of the formation of the Earth; reason being that the Earth is so structured, ordered, and organized to be able to sustain life forms. I will take some time to delve into the orderliness of the Earth when talking about the "greenhouse effect."

Regardless of the exact process of its formation, the Earth has since undergone significant changes and is now a diverse and dynamic planet with a rich history of life.

From a religious perspective, the creation of the Earth varies depending on the particular faith. Here are a few examples:

Christianity: In the biblical account of creation in Genesis Chapter 1, God created the Earth and everything in it in six days, as described in the book of Genesis. According to this account, God created light on the first day, separated the waters to form the skies and the land on the second day, and created plants and trees on the third day. On the fourth day, God created the sun, moon, and stars. On the fifth day, God created the sea creatures and birds, and on the sixth day, God created the land animals and humans. According to the biblical book, a day in the eyes of God is 1000 years in the calendar of humans. So, literally, God would have taken 6000 years to create the Earth and the universe as a whole. Each day, 1000 years may signify a process of time and formation.

Genesis Chapter 1:28 also made mention of God's instruction to humans after he finished creating them:

"Then God blessed them, and God said to them, 'Be fruitful and multiply; fill the earth and subdue it; have dominion over the fish of the sea, over the birds of the air, and over every living thing that moves on the earth.'"

Islam: According to Islamic belief, Allah created the universe and everything in it, including the Earth, in six days. This account is described in the Qur'an, the Islamic holy book.

Hinduism: Hinduism teaches that the universe was created, destroyed, and re-created again and again in a cyclical process. The creation of the Earth and everything on it is described in various Hindu texts, including the Puranas and the Vedas.

Judaism: In Jewish tradition, the creation of the Earth is described in the book of Genesis. According to this account, God created the world in six days and rested on the seventh.

These are just a few examples of how different religions view the creation of the Earth. It's important to note that religious beliefs about the creation of the Earth are not scientifically verifiable and should be understood within the context of religious faith.

In my own perspective, which seems to align more with the religious belief (even though I consider myself more spiritual than religious), I believe the complex, organized, and orderly nature of the Earth was put together by an infinite intelligence, whom I refer to as God Almighty.

The force of gravity between the Earth and the sun keeps the Earth in its orbit and stops it from flying off into space. Gravity is a fundamental force of nature that arises from the mass and energy of objects. The more massive an object is, the more gravity it generates. The strength of the gravitational force between two objects decreases with distance, but it never truly goes away. A higher intelligence is definitely responsible for keeping the Earth hanging in space and orbiting around the sun at a distance of about 93 million miles (149.6 million km).

The Greenhouse Effect

The greenhouse effect is a process that occurs when gases in Earth's atmosphere trap the sun's heat. This process makes Earth much warmer than it would be without an atmosphere. The greenhouse effect is one of the things that makes Earth a comfortable place to live.

How does the greenhouse effect work?

As you might expect from the name, the greenhouse effect works… like a greenhouse! A greenhouse is a building with glass walls and a glass roof. Greenhouses are used to grow plants, such as tomatoes and tropical flowers.

A greenhouse stays warm inside, even during the winter. In the daytime, sunlight shines into the greenhouse and warms the plants and air inside. At nighttime, it's colder outside, but the greenhouse stays pretty warm inside. That's because the glass walls of the greenhouse trap the sun's heat.

The greenhouse effect works much the same way on Earth. Gases in the atmosphere, such as carbon dioxide, trap heat similar to the glass roof of a greenhouse. These heat-trapping gases are called greenhouse gases.

During the day, the sun shines through the atmosphere. Earth's surface warms up in the sunlight. At night, Earth's surface cools, releasing heat back into the air. But some of the heat is trapped by the greenhouse gases in the atmosphere. That's what keeps our Earth a warm and cozy 58 degrees Fahrenheit (14 degrees Celsius), on average.

How are humans impacting the greenhouse effect?

Human activities are changing Earth's natural greenhouse effect. Burning fossil fuels like coal and oil puts more carbon dioxide into our atmosphere.

NASA has observed increases in the amount of carbon dioxide and some other greenhouse gases in our atmosphere. Too much of these greenhouse gases can cause Earth's atmosphere to trap more and more heat. This causes Earth to warm up.

What reduces the greenhouse effect on Earth?

Just like a glass greenhouse, Earth's greenhouse is also full of plants! Plants can help to balance the greenhouse effect on Earth. All plants—from giant trees to tiny phytoplankton in the ocean—take in carbon dioxide and give off oxygen.

The ocean also absorbs a lot of excess carbon dioxide from the air. Unfortunately, the increased carbon dioxide in the ocean changes the water, making it more acidic. This is called ocean acidification.

More acidic water can be harmful to many ocean creatures, such as certain shellfish and coral. Warming oceans—from too many greenhouse gases

gewo-intl.org/the-threat-to-earth/

in the atmosphere—can also be harmful to these organisms. Warmer waters are a main cause of coral bleaching.

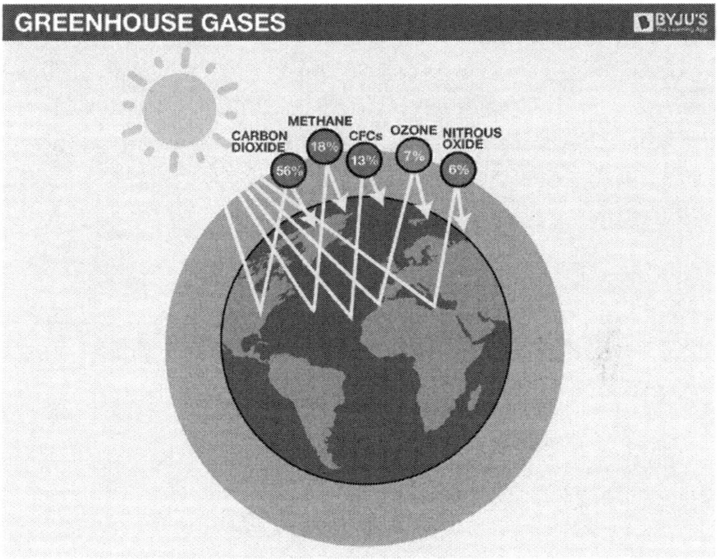

The diagram shows that greenhouse gases such as carbon dioxide are the primary cause for the greenhouse effect.

(Source: https://byjus.com/biology/greenhouse-effect-gases/)

Causes of Greenhouse Effect

The major causes of the greenhouse effect are:

Burning of Fossil Fuels

Fossil fuels are an important part of our lives. They are widely used in transportation and to produce electricity. Burning of fossil fuels releases carbon dioxide. With the increase in population, the utilization of fossil fuels has increased. This has led to an increase in the release of greenhouse gases in the atmosphere.

Deforestation

Plants and trees take in carbon dioxide and release oxygen. Due to the cutting of trees, there is a considerable increase in the greenhouse gases, which increases the Earth's temperature.

Farming

Nitrous oxide used in fertilizers is one of the contributors to the greenhouse effect in the atmosphere.

Industrial Waste and Landfills

The industries and factories produce harmful gases that are released into the atmosphere.

Landfills also release carbon dioxide and methane, which adds to the greenhouse gases.

Effects of Greenhouse Gases

The main effects of increased greenhouse gases are:

Global Warming

It is the phenomenon of a gradual increase in the average temperature of the Earth's atmosphere. The main cause for this environmental issue is the increased volumes of greenhouse gases, such as carbon dioxide and methane, released by the burning of fossil fuels, emissions from vehicles, industries, and other human activities.

Depletion of the Ozone Layer
The ozone layer protects the Earth from harmful ultraviolet rays from the sun. It is found in the upper regions of the stratosphere. The depletion of the ozone layer results in the entry of the harmful UV rays to the Earth's surface, which might lead to skin cancer and can also change the climate drastically.

The major cause of this phenomenon is the accumulation of natural greenhouse gases, including chlorofluorocarbons, carbon dioxide, methane, etc.

Smog and Air Pollution
Smog is formed by the combination of smoke and fog. It can be caused both by natural means and man-made activities.

In general, smog is generally formed by the accumulation of more greenhouse gases, including nitrogen and sulfur oxides. The major contributors to the formation of smog are automobile and industrial emissions, agricultural fires, natural forest fires, and the reaction of these chemicals among themselves.

Acidification of Water Bodies
Increase in the total amount of greenhouse gases in the air has turned most of the world's water bodies acidic. The greenhouse gases mix with the rainwater and fall as acid rain. This leads to the acidification of water bodies.

Also, the rainwater carries the contaminants along with it and falls into the river, streams, and lakes, thereby causing their acidification.

Runaway Greenhouse Effect

This phenomenon occurs when the planet absorbs more radiation than it can radiate back. Thus, the heat lost from the Earth's surface is less and the temperature of the planet keeps rising. Scientists believe that this phenomenon took place on the surface of Venus billions of years ago.

This phenomenon is believed to have occurred in the following manner:

A runaway greenhouse effect arises when the temperature of a planet rises to a level of the boiling point of water. As a result, all the water from the oceans converts into water vapour, which traps more heat coming from the sun and further increases the planet's temperature. This eventually accelerates the greenhouse effect. This is also called the "positive feedback loop."

There is another scenario giving way to the runaway greenhouse effect. Suppose the temperature rise, due to the above causes, reaches such a high level that the chemical reactions begin to occur. These chemical reactions drive carbon dioxide from the rocks into the atmosphere. This would heat the surface of the planet, which would further accelerate the transfer of carbon dioxide from the rocks to the atmosphere, giving rise to the runaway greenhouse effect.

In simple words, increasing the greenhouse effect gives rise to a runaway greenhouse effect, which would increase the temperature of the Earth to such an extent that no life would exist in the near future.

Definition of Climate Change and Its Causes

Climate change refers to the long-term and persistent shifts in global or regional climate patterns, including changes in temperature, precipitation, and sea levels, that occur over decades or longer periods of time. It is largely driven by human activities, especially the burning of fossil fuels, deforestation, and industrial processes, which release large amounts of greenhouse gases (GHGs) such as carbon dioxide, methane, and nitrous oxide into the atmosphere.

Greenhouse gases trap heat in the Earth's atmosphere, acting like a blanket around the planet and keeping it warm. However, the increased concentration of these gases in the atmosphere from human activities is causing the Earth's temperature to rise, resulting in a range of impacts on the environment and society, including more frequent and severe heatwaves, droughts, floods, storms, and sea level rise. Climate change also has other indirect effects such as changes in ecosystems, agriculture, and water availability, as well as impacts on human health, migration, and social and economic systems.

The primary human activities that contribute to climate change include the burning of fossil fuels (coal, oil, and natural gas) for energy and transportation, deforestation, agricultural practices such as livestock farming and fertilizer use, and industrial processes. These activities release large amounts of carbon dioxide and other GHGs into the atmosphere, which trap heat and cause global temperatures to rise.

In addition to human activities, natural factors such as solar activity, volcanic eruptions, and changes in the Earth's orbit also affect the climate. However, these natural factors are not sufficient to explain the current rapid pace of climate change observed in recent decades.

To address climate change, it is necessary to reduce greenhouse gas emissions through a range of measures, including transitioning to renewable energy sources, improving energy efficiency, reducing deforestation, and changing agricultural practices. The international community has also developed various agreements and frameworks, such as the Paris Agreement, to combat climate change and mitigate its impacts.

Brief History of Climate Change and Its Recognition as a Global Threat

Climate change has been a natural process on Earth for millions of years, but the current and rapid pace of climate change is largely caused by human activities that have intensified since the Industrial Revolution in the 18th century.

In the mid-1800s, scientists such as Joseph Fourier and John Tyndall observed the greenhouse effect and how it contributed to the Earth's temperature, as greenhouse gases trap heat in the atmosphere. In 1896, Svante Arrhenius, a Swedish chemist, calculated that a doubling of carbon dioxide in the atmosphere would lead to a significant increase in global temperature, an early recognition of the potential impact of human activities on the climate.

In the 20th century, researchers and organizations began to take notice of the effects of climate change on the environment and society. In the 1950s, scientists began to measure carbon dioxide levels in the atmosphere and recognized the significant increase in concentrations caused by human activities.

In 1988, the United Nations established the Intergovernmental Panel on Climate Change (IPCC), an international organization that assesses the science, impacts, and mitigation of climate change. The IPCC has issued several assessment reports that have provided compelling evidence of the reality and seriousness of climate change, including the role of human activities in causing it.

Over the years, global leaders have recognized climate change as a serious threat to the environment, economies, and societies worldwide. In 1992, the United Nations Framework Convention on Climate Change (UNFCCC) was signed by countries around the world, with the objective of stabilizing greenhouse gas concentrations in the atmosphere and preventing dangerous human interference with the climate system. The

gewo-intl.org/the-threat-to-earth/

Paris Agreement, signed in 2015, is the most significant global agreement to date, with the goal of limiting global temperature rise to well below 2 degrees Celsius above pre-industrial levels, and pursuing efforts to limit the temperature increase to 1.5 degrees Celsius.

In recent years, climate change has gained increasing recognition as a global threat, with more countries and organizations taking action to reduce greenhouse gas emissions, protect the environment, and mitigate the impacts of climate change on societies and economies. Overall, the increasing recognition of climate change as a global threat has been driven by a combination of scientific evidence, policy action, and public awareness, and has led to increased action to reduce greenhouse gas emissions and mitigate the impacts of climate change on societies and economies.

Overview of the Impact of Climate Change on Humanity and the Planet

Climate change is one of the most pressing global challenges of our time, with significant impacts on both the planet and humanity. The following is a detailed overview of the impact of climate change on the environment, societies, and economies:

Environmental Impacts:

Rising Temperatures: Climate change is causing the Earth's temperature to rise, which is leading to melting ice caps, rising sea levels, and increasing heatwaves, droughts, and wildfires.

Extreme Weather: Climate change is making extreme weather events more frequent and severe, such as hurricanes, typhoons, floods, and heatwaves, which can cause significant damage to the environment and human communities.

Ocean Acidification: The ocean absorbs much of the excess carbon dioxide that is released into the atmosphere, which is leading to increased acidity levels and damage to marine ecosystems.

Biodiversity Loss: Climate change is causing significant losses in biodiversity, with many species being threatened or endangered by changes in their habitats and food sources.

Social Impacts:

Health: Climate change is leading to an increase in the frequency and intensity of heatwaves, which can cause heat stroke, dehydration, and other health issues. It can also lead to the spread of diseases and other health impacts.

Water Availability: Climate change is leading to changes in water availability and quality, which can impact agriculture, food security, and access to drinking water.

Migration: Climate change is leading to displacement and migration of people, particularly in low-lying coastal areas and regions affected by drought and other environmental impacts.

Conflict: Climate change is leading to increased competition over resources, such as water and food, which can lead to conflicts between communities and nations.

Economic Impacts:

Loss of Productivity: Climate change is leading to disruptions in supply chains and transportation, as well as damage to infrastructure, which can lead to decreased productivity and economic output.

gewo-intl.org/the-threat-to-earth/

Economic Disruption: Climate change is causing disruptions in financial markets, as well as increased insurance costs and losses from extreme weather events, which can impact economies worldwide.

Cost of Adaptation: Adapting to climate change can be costly, and the cost of inaction is even higher. Countries and businesses must invest in adaptation and mitigation measures to address climate change and its impacts.

Inequality: Climate change impacts are not distributed equally, with the poorest and most vulnerable communities often being the most affected. This can exacerbate social and economic inequalities.

Overall, climate change is having significant impacts on the environment, societies, and economies worldwide, and urgent action is needed to reduce greenhouse gas emissions, adapt to the changing climate, and mitigate its impacts on humanity and the planet.

Chapter 2

Climate Change and Its Effects on the Natural World

Rising Global Temperatures and Their Effect on Wildlife, Ecosystems, and Biodiversity

Rising global temperatures, largely caused by human activities such as the burning of fossil fuels, have a profound effect on wildlife, ecosystems, and biodiversity. As the temperature continues to rise, there are increasing concerns about the irreversible damage to our planet's natural systems and the survival of the species that rely on them.

The impacts of rising temperatures on wildlife and ecosystems are numerous and far-reaching. Some of the most notable effects include changes in migration patterns, habitat loss, and altered breeding cycles. For example, as the Arctic sea ice continues to melt due to rising temperatures, it has forced some polar bears to swim longer distances in search of food, leading to a decline in their population. Similarly, many migratory bird species are arriving at their breeding grounds earlier in the year, but their food sources have not yet become available, resulting in a decline in their population.

Rising temperatures are also causing changes in the timing and extent of seasonal cycles, leading to mismatches between species and their food sources. This can result in declines in populations, as well as changes in the distribution and abundance of certain species. For example, changes in precipitation patterns and temperature have led to the decline of amphibian populations in many areas, as their breeding and development cycles are dependent on consistent water availability.

Additionally, climate change is leading to the loss of habitats as species are forced to move to areas with more favorable conditions. This displacement often causes conflict between species and can lead to changes in the composition of entire ecosystems. In many cases, these changes can have cascading effects on the rest of the ecosystem, leading to a decline in biodiversity and a loss of ecosystem services.

Overall, the effects of rising global temperatures on wildlife, ecosystems, and biodiversity are complex and far-reaching. It is important that we take steps to mitigate these impacts by reducing our greenhouse gas emissions, protecting important habitats, and implementing adaptive management strategies. By doing so, we can help to ensure the survival of the planet's most vulnerable species and preserve the ecological systems that sustain us all.

Melting of Polar Ice Caps and Rising Sea Levels

Melting of polar ice caps, caused by global warming and climate change, is a major environmental concern that has the potential to cause devastating impacts on the planet. The polar ice caps of the Arctic and Antarctica contain massive amounts of ice, and as they melt, they contribute to rising sea levels, which could lead to flooding, coastal erosion, and destruction of habitats for various marine species.

The Arctic region has been experiencing rapid melting of its sea ice, which is projected to continue with further global warming. This is due to the fact that the Arctic is warming at a faster rate than the rest of the planet, resulting in the melting of the sea ice that covers the Arctic Ocean. The Arctic sea ice has decreased by around 13% per decade since the

1980s, with the area of the ice cover being the lowest it has been in recorded history.

The melting of the Arctic sea ice has numerous consequences. One of the major concerns is the rising sea levels, which could lead to coastal flooding and erosion in many regions around the world. This could cause significant economic and social impacts, as well as a loss of habitats for marine species that depend on coastal environments. Additionally, the melting of the sea ice has significant implications for the global climate, as the Arctic acts as a "refrigerator" for the planet, helping to regulate the temperature and weather patterns.

The melting of the Antarctic ice sheet is also a significant concern. While the Arctic sea ice is floating on the ocean, the Antarctic ice sheet is a land-based ice sheet, and it contains an estimated 90% of the world's ice. If the entire Antarctic ice sheet were to melt, it could lead to a rise in sea levels of up to 60 meters (196 feet). While the complete melting of the Antarctic ice sheet is unlikely in the near future, its melting has already led to significant changes in the global sea level.

One of the major consequences of the melting of the Antarctic ice sheet is the impact it has on the ocean currents. The melting of the ice sheet has caused an influx of fresh water into the Southern Ocean, which can disrupt the ocean currents that regulate the Earth's climate. This can cause changes in the weather patterns and ocean temperatures, which can affect marine ecosystems and fisheries.

In conclusion, the melting of the polar ice caps, particularly in the Arctic and Antarctic, is a major environmental concern that has significant implications for the planet. The rise in sea levels, which is caused by the melting of the ice, has the potential to cause major impacts on coastal regions, ecosystems, and human communities. It is important that we take steps to mitigate the effects of global warming and climate change by reducing greenhouse gas emissions and implementing sustainable policies to protect our planet's natural systems.

Extreme Weather Events and Their Impact on Agriculture and Food Security

Climate change is causing more frequent and severe extreme weather events, which have significant impacts on agriculture and food security. The increasing global temperature is leading to changes in the Earth's weather patterns, such as more intense heat waves, droughts, floods, and hurricanes, which are having significant impacts on crops and agricultural production.

Heatwaves are becoming more frequent and intense due to climate change, leading to higher temperatures and longer periods of hot weather. This can cause heat stress in crops, leading to lower yields and lower quality of crops. For example, in 2010, a heat wave in Russia caused widespread crop losses, leading to a decrease in food exports and higher food prices. In 2018, Europe experienced an extended heatwave that caused significant crop losses, particularly in wheat and barley, leading to increased food prices and potential food shortages.

Droughts are also becoming more severe and more frequent due to climate change, leading to reduced water availability for crops and livestock. This can cause lower crop yields, loss of livestock, and food shortages. For example, in 2012, the United States experienced a severe drought that caused the prices of corn and soybeans to rise significantly, which led to food price spikes in many countries around the world. In Southern Africa, droughts have become more frequent and severe, leading to food insecurity and malnutrition.

Floods are also becoming more frequent and more severe due to climate change, leading to damage to crops and agricultural infrastructure. Floods can lead to soil erosion and loss of topsoil, which reduces soil fertility and can cause lower crop yields. For example, in 2010, severe floods in Pakistan destroyed crops and agricultural infrastructure, leading to a decline in food availability and increased food prices. In 2011, Thailand experienced severe floods that damaged rice fields and disrupted food production, leading to a global rice shortage.

Hurricanes and tropical storms are also becoming more frequent and more severe due to climate change, leading to damage to crops and agricultural infrastructure. Hurricanes and tropical storms can cause damage to crops and can also disrupt food distribution systems, leading to food shortages. For example, in 2017, Hurricane Harvey caused significant damage to Texas and Louisiana, leading to crop losses and disruption of food production. Still, in 2017, Hurricane Maria devastated Puerto Rico, causing widespread damage to agricultural land, crops, and infrastructure, leading to a food crisis on the island.

Extreme weather events also have long-term impacts on agriculture and food security. For example, repeated droughts can lead to land degradation, which can reduce the productivity of agricultural land. This can lead to long-term food insecurity in regions that rely on agriculture as a primary source of food and income.

In conclusion, climate change is contributing to more frequent and severe extreme weather events, which have significant impacts on agriculture and food security. Heatwaves, droughts, floods, and hurricanes are causing significant damage to crops and agricultural infrastructure, leading to reduced yields, lower quality of crops, and potential food shortages. It is important to implement sustainable agricultural practices and develop strategies to adapt to the impacts of climate change to ensure food security for all.

Climate-Driven Changes in Ocean Chemistry and Its Effect on Marine Life

Climate-driven changes in ocean chemistry, including changes in temperature, salinity, and acidity, can have significant effects on marine life. These changes are largely caused by the absorption of excess atmospheric carbon dioxide by the ocean, which can lead to a process known as ocean acidification.

Ocean acidification occurs when carbon dioxide dissolves in seawater, forming carbonic acid. This process increases the acidity of the ocean, making it more difficult for marine organisms to form and maintain their shells and skeletons, which are made of calcium carbonate. As a result, marine organisms such as corals, mollusks, and plankton may struggle to survive and reproduce.

In addition to ocean acidification, rising ocean temperatures can also have significant effects on marine life. Warmer waters can lead to the expansion of oxygen-poor areas, which can suffocate marine life, as well as the proliferation of harmful algal blooms and other toxic organisms that can cause illness or death to marine life. For example, in 2018, a massive toxic algal bloom in Florida's Gulf of Mexico caused a state of emergency and resulted in the deaths of dolphins, manatees, and sea turtles.

Climate-driven changes in ocean chemistry can also have cascading effects throughout marine ecosystems. For example, changes in the abundance of phytoplankton, which are the base of the marine food chain, can have profound effects on the entire ecosystem. Phytoplankton rely on dissolved carbon dioxide to grow, and as the ocean becomes more acidic, their growth may be inhibited, leading to changes in the availability of food for other marine organisms.

In addition, ocean acidification can alter the behavior and physiology of marine organisms. For example, some studies suggest that acidification can impair the sense of smell in some species of fish, making it more difficult for them to find food or avoid predators. Other studies have suggested that acidification can impair the ability of some species to navigate, potentially leading to declines in populations and changes in species distribution.

gewo-intl.org/the-threat-to-earth/

Overall, climate-driven changes in ocean chemistry can have significant and far-reaching effects on marine life and, ultimately, on human society. As the ocean continues to absorb carbon dioxide from the atmosphere, it is crucial that we take action to reduce greenhouse gas emissions and mitigate the impacts of climate change on marine ecosystems.

Chapter 3

Climate Change and Its Impact on Human Health

Increased Air Pollution and Its Effect on Respiratory and Cardiovascular Health

Climate change can exacerbate air pollution by altering atmospheric conditions and increasing the frequency and intensity of extreme weather events. This can have significant effects on respiratory and cardiovascular health, particularly in vulnerable populations such as children, the elderly, and those with pre-existing health conditions.

Climate change can affect air pollution in several ways. For example, higher temperatures can lead to the formation of more ground-level ozone, a key component of smog that can cause respiratory and cardiovascular problems. In addition, more frequent wildfires, which are becoming more common in many parts of the world due to climate change, can release large amounts of particulate matter and other pollutants into the air. Finally, extreme weather events such as hurricanes, tornadoes, and dust storms can also cause spikes in air pollution levels.

The health effects of air pollution can be severe. Exposure to air pollution has been linked to a range of respiratory and cardiovascular problems, including asthma, chronic obstructive pulmonary disease (COPD), lung cancer, heart disease, stroke, and premature death. Children and the elderly are particularly vulnerable, as are those with pre-existing respiratory or cardiovascular conditions.

For example, a study published in the journal *Lancet Planetary Health*, in 2020, found that exposure to air pollution was responsible for over 160,000 premature deaths in the United States in 2018. Another study, published in the *New England Journal of Medicine*, in 2019, estimated that air pollution was responsible for over 8 million premature deaths worldwide in 2015.

The health effects of air pollution can be particularly pronounced in low-income communities and communities of color, which are often located near sources of pollution such as highways and industrial facilities. These communities may also be more vulnerable to the effects of extreme weather events, which can exacerbate existing health disparities.

In addition to the direct health effects of air pollution, it can also have indirect effects on human health. For example, air pollution can reduce crop yields and impair the nutritional value of crops, which can have negative effects on human health. Air pollution can also have economic effects, such as increased healthcare costs and lost productivity due to illness.

The link between climate change and air pollution underscores the urgent need to reduce greenhouse gas emissions and take action to mitigate the impacts of climate change on human health. This includes promoting clean energy sources, improving public transportation, and implementing policies to reduce air pollution from industrial sources and other polluting activities.

gewo-intl.org/the-threat-to-earth/

Spread of Disease Through Insects and Pests in Warmer Climates

Climate change is causing shifts in the distribution and behavior of insects and other pests, leading to an increased risk of the spread of infectious diseases. Here are some examples of how climate change is affecting the spread of disease through insects and pests:

Mosquito-borne diseases: Mosquitoes are known to transmit several serious diseases, including malaria, dengue fever, and Zika virus. Warmer temperatures and increased rainfall associated with climate change are contributing to the expansion of mosquito populations in many parts of the world, increasing the risk of mosquito-borne diseases. For example, in 2019, the World Health Organization reported a surge in dengue fever cases in many countries, including Bangladesh, Brazil, and the Philippines, partly due to the effects of climate change.

Tick-borne diseases: Ticks are known to transmit several diseases, including Lyme disease and Rocky Mountain spotted fever. As temperatures warm, ticks are expanding their range into new areas and becoming active for longer periods of time, leading to an increased risk of tick-borne diseases. A study published in the journal *Nature Communications,* in 2018, found that the number of people at risk of tick-borne diseases in the United States had more than doubled over the previous 20 years.

Rodent-borne diseases: Rodents such as rats and mice can transmit several serious diseases, including hantavirus and leptospirosis. Warmer temperatures and changes in precipitation patterns associated with climate change can lead to an increase in the availability of food and water for rodents, leading to an increase in rodent populations and an increased risk of rodent-borne diseases. A study published in the journal *Environmental Health Perspectives,* in 2019, found that climate change was contributing to an increase in rodent populations in New York City, increasing the risk of rodent-borne diseases in the city's human population.

Plant pests and diseases: Climate change is affecting the distribution and behavior of many plant pests and diseases, leading to an increased risk of crop loss and food insecurity. For example, in 2020, locust swarms devastated crops across several countries in East Africa, including Kenya, Ethiopia, and Somalia. The swarms were fueled in part by unusually heavy rains, which provided ideal breeding conditions for the locusts.

These are just a few examples of the many ways in which climate change is affecting the spread of disease through insects and pests. The impacts of these changes on human health are likely to be significant, particularly in vulnerable populations in low- and middle-income countries. Addressing climate change and taking steps to control the spread of infectious diseases will be crucial for protecting human health in the years to come.

Contaminated Water Sources and Its Impact on Public Health

Climate change can lead to changes in the quantity and quality of water resources, which can in turn affect the availability and safety of drinking water. Here are some examples of the impact of contaminated water sources on public health as a result of climate change:

Waterborne diseases: Changes in precipitation patterns and temperatures associated with climate change can lead to changes in the quality of water resources, increasing the risk of waterborne diseases. For example, heavy rainfall can cause sewage overflows and runoff from farms and other sources to enter waterways, leading to an increased risk of diseases such as cholera, typhoid fever, and hepatitis A. In 2018, an outbreak of Legionnaires' disease in New York City was linked to contaminated cooling towers, which had become a breeding ground for

Legionella bacteria due to the warm and humid conditions associated with climate change.

Foodborne illnesses: Changes in water availability and quality can also affect agriculture and the safety of the food supply. For example, droughts can reduce the availability of water for irrigation and increase the risk of foodborne illnesses from contaminated water used to grow crops. In 2020, an outbreak of E. coli infections in the United States was linked to contaminated lettuce grown in an area affected by drought and poor water quality.

Chemical contamination: Changes in water availability and quality can also increase the risk of chemical contamination of water sources, which can have serious health impacts. For example, droughts can lead to the concentration of pollutants in water sources due to reduced dilution, and increased temperatures can increase the release of chemicals from soils and sediments. In 2016, the city of Flint, Michigan experienced a public health crisis when high levels of lead were found in the city's water supply, due in part to changes in water sources and treatment methods associated with drought conditions.

Mental health: Climate-related changes in water resources can also have impacts on mental health. For example, prolonged droughts can lead to stress and anxiety for farmers and others who rely on water resources for their livelihoods, while flooding and other extreme weather events can lead to displacement and trauma.

These are just a few examples of the ways in which contaminated water sources, as a result of climate change, can impact public health. Addressing the impacts of climate change on water resources and taking steps to ensure safe and secure access to clean water will be crucial for protecting public health in the years to come.

Displacement of People Due to Extreme Weather Events and Rising Sea Levels

Climate change is causing extreme weather events and rising sea levels, which can displace people from their homes and communities. Here are some examples of the displacement of people as a result of these impacts of climate change:

Extreme weather events: Climate change is causing more frequent and intense extreme weather events such as hurricanes, floods, and wildfires. These events can cause significant damage to homes and infrastructure, making it unsafe for people to remain in affected areas. For example, in 2017, Hurricane Harvey caused massive flooding in Houston, Texas, leading to the displacement of more than 30,000 people from their homes. In 2018, wildfires in California destroyed thousands of homes, forcing tens of thousands of people to evacuate.

Sea level rise: As temperatures continue to rise, glaciers and ice sheets are melting, causing sea levels to rise. This rise in sea level is a slow process, but it can have significant impacts on low-lying coastal areas over time. For example, in the Pacific island nation of Kiribati, rising sea levels have caused saltwater to contaminate freshwater sources, making it difficult for people to find safe drinking water. The government of Kiribati has even purchased land in Fiji as a potential future home for its citizens if rising sea levels make the island uninhabitable.

Desertification: Climate change can also cause areas to become too dry to support agriculture and other forms of life, a process known as desertification. This can force people to leave their homes in search of more hospitable environments. For example, in parts of sub-Saharan Africa, changing rainfall patterns have led to droughts and desertification, causing widespread food shortages and displacement.

Conflict: The displacement of people as a result of climate change can also contribute to conflict and political instability. As people are forced to migrate from their homes and compete for resources, tensions can rise between communities, and conflicts can erupt. For example, in Syria, a drought that lasted from 2006 to 2011 contributed to food shortages and displacement, which may have played a role in the outbreak of civil war.

These are just a few examples of the displacement of people as a result of climate change. The impacts of climate change are expected to become more severe in the coming years, making it essential for governments and communities to prepare and adapt to the changes ahead to minimize the impact on human lives.

Chapter 4

Climate Change and Its Effect on the Global Economy

Loss of Coastal Infrastructure and Properties

Climate change is causing a number of impacts on coastal infrastructure and properties, including loss of land, damage to buildings and infrastructure, and increased flooding and erosion. Here are some of the specific ways in which climate change is affecting coastal infrastructure and properties:

Sea level rise: One of the most obvious impacts of climate change on coastal infrastructure and properties is sea level rise. As global temperatures increase, glaciers and ice caps melt, causing the sea level to rise. This can lead to flooding of coastal areas, erosion of beaches and shorelines, and damage to buildings and infrastructure. For example, in the United States, coastal areas are experiencing an average sea level rise of about 3.2 millimeters per year, which is causing increased flooding and erosion.

Storm surges: As sea levels rise, storms can cause more damage to coastal infrastructure and properties. Storm surges, which are the walls of water that form during hurricanes and other severe weather events, can cause flooding and damage to buildings and infrastructure. For example, Hurricane Katrina caused widespread damage to coastal infrastructure and properties in the Gulf of Mexico region in 2005.

Erosion: Climate change is also causing increased erosion of coastlines, which can lead to loss of land and damage to buildings and infrastructure. Erosion is caused by a combination of rising sea levels, increased storm activity, and human activities such as coastal development and sand mining. For example, the erosion of the coastline in Miami Beach, Florida, has caused the loss of beaches and threatened buildings and infrastructure.

Ocean acidification: Another impact of climate change on coastal infrastructure and properties is ocean acidification. As carbon dioxide levels in the atmosphere increase, the oceans absorb more of this gas, causing the pH level of the water to decrease. This can harm marine life and coral reefs, which are important for protecting coastlines from erosion and storm surges. For example, the Great Barrier Reef in Australia is under threat from ocean acidification, which could lead to the loss of this important natural barrier.

Heatwáves: Climate change is also causing more frequent and intense heatwaves, which can damage buildings and infrastructure. Extreme heat can cause materials to expand and contract, leading to cracking and other forms of damage. For example, in 2018, a heatwave in the United Kingdom caused a number of train tracks to buckle, leading to widespread disruption of rail services.

The loss of coastal infrastructure and properties as a result of climate change is a serious concern for many coastal communities around the world. In addition to the impacts listed above, climate change is also causing increased saltwater intrusion into freshwater supplies, which can harm agriculture and other industries. It is important for governments and communities to take action to adapt to these impacts and mitigate the causes of climate change to minimize future losses.

Decline in Agricultural Production and Food Security

Climate change is contributing to the decline in agricultural production and food security in several ways. Here are some of the main ways in which climate change is affecting agriculture and food security:

Changes in temperature and precipitation: Climate change is causing changes in temperature and precipitation patterns that are making it harder for farmers to produce crops. Rising temperatures are causing more frequent heatwaves and droughts, which can lead to crop failure and reduced yields. Changes in precipitation patterns are causing more frequent and intense floods and droughts, which can also damage crops and reduce yields. For example, in sub-Saharan Africa, changing rainfall patterns have made it more difficult for farmers to plan their planting and harvesting cycles, leading to reduced yields.

Changes in pest and disease patterns: Climate change is also causing changes in the patterns of pests and diseases that can harm crops. Warmer temperatures can lead to the spread of pests and diseases into new areas, while changes in precipitation patterns can create conditions that favor the growth of certain pests and diseases. For example, in East Africa, rising temperatures are causing the spread of maize lethal necrosis disease, which can destroy entire crops.

Water scarcity: Climate change is causing increased water scarcity in many regions, which is making it harder for farmers to irrigate their crops. In some regions, melting glaciers and reduced snowpack are reducing the amount of water available for irrigation, while in other regions, changes in precipitation patterns are making it harder for farmers to predict when and where they will have access to water. For example, in India, changing rainfall patterns are causing increased water scarcity in many regions, leading to reduced crop yields.

Soil degradation: Climate change is also contributing to soil degradation, which can reduce the fertility of the land and make it harder for farmers to produce crops. Rising temperatures and changing precipitation patterns can cause soil erosion and nutrient depletion, while increased pest and disease pressure can also harm the soil. For example, in the United States, changing precipitation patterns are causing increased erosion in many regions, leading to soil degradation and reduced crop yields.

Extreme weather events: Finally, climate change is causing more frequent and intense extreme weather events, such as hurricanes, floods, and wildfires, which can damage crops and reduce yields. Extreme weather events can destroy entire crops, damage infrastructure such as irrigation systems and storage facilities, and make it harder for farmers to access markets to sell their produce. For example, in 1998 in Central America, Hurricane Mitch destroyed entire coffee plantations and caused widespread food shortages.

The decline in agricultural production and food security is a serious concern for many regions around the world, especially in developing countries where agriculture is a key industry. It is important for governments and communities to take action to adapt to these impacts and mitigate the causes of climate change to minimize future losses. This can include investing in climate-smart agriculture practices, improving irrigation and water management systems, and developing new crop varieties that are more resilient to changing climatic conditions.

Impact on Tourism and Recreation Industries

Climate change has been identified as one of the most significant challenges facing the global community, with a wide range of impacts that affect numerous sectors and industries. The tourism and recreation industries are particularly vulnerable to the effects of climate change, as they are heavily reliant on natural resources and outdoor activities that can be easily disrupted by changes in weather patterns, sea level rise, and other climate-related phenomena. In this answer, we will discuss the various impacts of climate change on the tourism and recreation industries and provide examples of how these impacts are already being felt around the world.

Changes in weather patterns: Climate change is already causing changes in weather patterns, including more frequent and intense storms, heat waves, and droughts. These changes can have a significant impact on tourism and recreation, as they can disrupt travel plans, cause cancellations, and make outdoor activities unsafe or unpleasant. For example, in 2017, Hurricane Irma devastated the Caribbean islands, causing widespread damage to tourism infrastructure and forcing many resorts to close temporarily. Similarly, in 2019, wildfires in California led to the closure of many national parks and forced the cancellation of numerous outdoor events.

Sea level rise: As global temperatures continue to rise, so does the level of the world's oceans, which can have a significant impact on coastal tourism and recreation industries. In many areas, rising sea levels are already causing beach erosion, increased storm surge, and flooding, which can damage infrastructure and deter tourists from visiting. For example, in 2018, Hurricane Florence caused extensive damage to the North Carolina coast, including flooding and beach erosion, which affected tourism in the region for months afterwards.

Impacts on wildlife and natural resources: Climate change can also have a significant impact on wildlife and natural resources that are important for tourism and recreation, such as coral reefs, forests, and wildlife habitats. As temperatures rise and weather patterns shift, these ecosystems can be disrupted, leading to habitat loss, changes in migration patterns, and decreased biodiversity. For example, the Great Barrier Reef in Australia has suffered significant damage in recent years due to coral bleaching caused by rising ocean temperatures, which has had a significant impact on the region's tourism industry.

Health impacts: Climate change can also have significant impacts on human health, which can affect tourism and recreation industries. For example, higher temperatures and more frequent heatwaves can pose health risks to tourists and outdoor enthusiasts, particularly those who are elderly or have pre-existing health conditions. In addition, increased air pollution and the spread of diseases such as malaria and dengue fever can also have a negative impact on tourism and recreation. For example, in 2017, air pollution in Delhi, India reached hazardous levels, leading many tourists to cancel their trips to the region.

The impacts of climate change on tourism and recreation industries are already being felt around the world and are expected to become more severe in the coming years. However, there are steps that can be taken to mitigate these impacts, such as investing in climate-resilient infrastructure, promoting sustainable tourism practices, and working to reduce greenhouse gas emissions. By taking action now, we can help ensure that these industries continue to thrive and provide economic benefits to communities around the world.

Increase in Energy Costs and Impact on the Energy Sector

Climate change has significant impacts on the energy sector, one of which is the increase in energy costs. The increase in energy costs is largely due to the implementation of policies and regulations aimed at reducing greenhouse gas emissions and mitigating the effects of climate change. In this answer, we will discuss in detail the impact of climate change on the energy sector's increase in energy costs.

Increase in costs of fossil fuel production: The production of fossil fuels such as coal, oil, and natural gas is a significant contributor to greenhouse gas emissions. As a result, many countries have implemented policies and regulations that impose taxes and fees on the production of these fuels. These policies increase the cost of producing and extracting fossil fuels, making them less competitive with renewable energy sources.

Higher costs of energy transportation: The transportation of fossil fuels, such as crude oil, is another significant contributor to greenhouse gas emissions. As a result, governments have implemented policies that require the use of more energy-efficient transportation methods, such as pipelines and electric trucks. These policies increase the cost of transporting fossil fuels, making them less competitive with renewable energy sources.

Investment in renewable energy sources: The implementation of policies and regulations aimed at reducing greenhouse gas emissions has led to an increase in the investment in renewable energy sources such as solar, wind, and hydropower. The increased investment in renewable energy has resulted in a decrease in the cost of production and installation of these energy sources, making them more cost-competitive with fossil fuels.

Demand for energy-efficient appliances and vehicles: Climate change policies have resulted in an increase in demand for energy-efficient appliances and vehicles. The higher demand for these products has led to the development of more energy-efficient technologies, such as LED lights, electric vehicles, and energy-efficient homes. The increased production and availability of these products have reduced energy consumption, leading to lower energy costs.

Incentives for energy efficiency: Many governments offer incentives for individuals and businesses that invest in energy-efficient technologies. These incentives include tax credits, rebates, and subsidies, which reduce the cost of investing in renewable energy and energy-efficient technologies.

Impact on vulnerable populations: The increase in energy costs resulting from climate change can have a significant impact on vulnerable populations, such as low-income households and those living in developing countries. These populations may be unable to afford the higher cost of energy, which can lead to energy insecurity and exacerbate existing health and social inequalities. As a result, policymakers must develop strategies to ensure that vulnerable populations have access to affordable and sustainable energy while promoting energy efficiency and renewable energy sources.

In conclusion, climate change policies and regulations have had significant impacts on the energy sector's increase in energy costs. The industry is investing in renewable energy and energy-efficient technologies, which are becoming more cost-competitive with traditional fossil fuels. However, policymakers and energy companies must consider the impact of higher energy costs on vulnerable populations and develop strategies to ensure that all populations have access to affordable and sustainable energy.

Chapter 5

Threat of Poor Nations to Global Fight Against Climate Change

The Leaking Boat Effect – How Poor Nations Will Capsize the Boat

The sailors on board the boat were sound asleep, lulled by the gentle rocking of the waves as the vessel made its way through the frigid waters. Suddenly, there was a loud, jarring crash that shook the boat violently, causing the sailors to be thrown from their bunks.

As they stumbled out of their cabins, still dazed and disoriented from the sudden impact, they quickly realized that their boat had collided with an iceberg looming ominously in the darkness. The once calm and peaceful sea had turned into a chaotic mess of crashing waves and scattered debris, and the sailors frantically scrambled to assess the damage and ensure that everyone was safe.

The sharp edges of the iceberg had torn through the hull of the big boat, and water was rapidly flooding in, threatening to capsize the vessel. With adrenaline pumping through their veins, the sailors worked to secure the damaged areas at both ends of the boat in order to keep the boat afloat.

As they labored through the night, bailing water and making repairs, the sailors couldn't help but reflect on how their journey began. It was a bright and sunny day when a rich merchant and a poor merchant set out on a long journey across the sea. They had been sailing for a few days when suddenly, dark clouds began to gather on the horizon, and the wind started to pick up. Within minutes, a terrible storm had descended upon them, and the boat was tossed around on the choppy waves like a toy.

The boat collided with a huge iceberg, and both ends of the boat started leaking. Despite their fear and the violent storm, the two merchants worked tirelessly to keep the boat afloat. The rich merchant, who had plenty of resources, quickly got to work mending his own side of the boat. Meanwhile, the poor merchant, who had no resources, struggled to mend his own side of the boat.

As the storm raged on, the water level in the boat began to rise rapidly. The poor merchant, despite his best efforts, could not keep up with the rate at which his side of the boat was filling up with water. The rich merchant, seeing that the poor merchant was struggling, refused to lend a hand, choosing instead to focus solely on his own survival.

As the water continued to fill up the boat, the poor merchant was eventually overcome and drowned. The rich merchant, despite having fixed his own side of the boat, was not spared. The water level continued to rise until the entire boat was submerged, and the rich merchant too drowned in the icy waters.

The tragic tale of the two merchants serves as a stark reminder of the dangers of neglecting the needs of the most vulnerable in times of crisis, especially when we live in an interconnected global village called the planet Earth. Just as the poor merchant was unable to save himself without assistance, the poorest nations in our world are often unable to adapt to the impacts of climate change without the support of the international community. Only by working together to address this global crisis can we hope to avoid a similar fate.

The point to learn from this scenario is that even when the rich nations are able to achieve net zero carbon emission by year 2050, the threat to

gewo-intl.org/the-threat-to-earth/

Earth as a result of global warming would still remain imminent. The poor nations who are unable to achieve net zero carbon emission reduction would continue to heat up the Earth's atmosphere, thereby contributing to the melting of polar ice, rising of water bodies, acidification of oceans, flooding and submerging of coastal cities, and finally the eventual overflooding of the Earth and the destruction of mankind.

Threat of Poor Nations to Global Fight Against Climate Change

Poor nations face several challenges in their fight against climate change. Here are some of the ways in which their situation threatens the global fight against climate change:

Limited resources: This refers to the lack of financial, technological, and natural resources available to poor nations to address the challenges posed by climate change. This lack of resources is a major obstacle to implementing effective climate change mitigation and adaptation strategies.

Financial resources: Poor nations often have limited financial resources to invest in climate change mitigation and adaptation measures. This can make it difficult for them to implement renewable energy projects, develop climate-resilient infrastructure, and invest in climate-smart agriculture. Without financial resources, it can be challenging for poor nations to reduce their greenhouse gas emissions and adapt to the impacts of climate change.

Bangladesh is a country in South Asia that is highly vulnerable to the impacts of climate change, such as flooding and cyclones. However, the country has limited financial resources to invest in climate change mitigation and adaptation measures, such as building sea walls or developing early warning systems.

Ethiopia is a country in East Africa that is highly vulnerable to the impacts of climate change, such as drought and food insecurity. However, the country has limited financial resources to invest in climate change mitigation and adaptation measures, such as developing drought-resistant crops or promoting sustainable land management practices.

Haiti is a country in the Caribbean that is highly vulnerable to the impacts of climate change, such as flooding and landslides. However, the country has limited financial resources to invest in climate change mitigation and adaptation measures, such as building infrastructure to withstand natural disasters or promoting reforestation.

Malawi is a country in Southern Africa that is highly vulnerable to the impacts of climate change, such as drought and food insecurity. However, the country has limited financial resources to invest in climate change mitigation and adaptation measures, such as developing early warning systems or promoting sustainable land management practices.

These examples demonstrate how poor nations often have limited financial resources to invest in climate change mitigation and adaptation measures, despite being highly vulnerable to its impacts. It is important for the global community to provide support and resources to help these nations build resilience to the impacts of climate change and transition to sustainable development pathways.

Technological resources: The lack of technological resources can make it difficult for poor nations to access and implement sustainable technologies that can reduce greenhouse gas emissions and enhance climate resilience. For example, poor nations may lack the technology necessary to implement renewable energy projects, develop drought-resistant crops, or improve water management practices. Without access to these technologies, it can be challenging for poor nations to address the challenges posed by climate change.

gewo-intl.org/the-threat-to-earth/

Examples of poor nations that are hard hit by the lack of technological resources are as follows:

Uganda is a country in East Africa that is facing a severe energy crisis. The country relies heavily on fossil fuels for energy, which are a significant contributor to greenhouse gas emissions. However, Uganda has limited access to renewable energy technologies, such as solar and wind power, which could help reduce its reliance on fossil fuels.

Afghanistan is a country in Central Asia that is facing a severe water crisis. The country has limited water resources and is highly vulnerable to drought and climate variability. However, Afghanistan has limited access to technology that could help improve water management practices, such as rainwater harvesting and drip irrigation.

Haiti is a country in the Caribbean that is facing a severe deforestation crisis. The country has lost much of its forest cover due to deforestation, which is a significant contributor to greenhouse gas emissions. However, Haiti has limited access to technology that could help reduce deforestation and promote reforestation, such as satellite imagery and remote sensing technology.

Sierra Leone is a country in West Africa that is facing a severe food security crisis. The country has limited arable land and is highly vulnerable to climate variability, such as drought and flooding. However, Sierra Leone has limited access to technology that could help improve crop yields and promote drought-resistant crops, such as improved seeds and irrigation technology.

These examples demonstrate how poor nations that lack technological resources face additional challenges in addressing the impacts of climate change. It is important for the global community to provide support and resources to help these nations access and implement sustainable technologies that can help reduce greenhouse gas emissions and enhance climate resilience.

Natural resources: Poor nations may also face limited natural resources, which can make it challenging for them to adapt to the impacts of climate change. For example, poor nations may have limited access to freshwater resources, making it difficult to address water scarcity and drought. They may also face limitations in arable land or access to marine resources, limiting their ability to sustainably manage natural resources and support their populations.

Limited resources are a significant challenge for poor nations in their fight against climate change. Addressing this challenge will require greater investment and support from the international community, including increased financial assistance, technology transfer, and capacity building. By addressing the limitations on resources, poor nations can better address the challenges posed by climate change and contribute to the global fight against climate change. Let's take a deep dive into poor nations that are hard hit by this challenge.

Yemen is a country in the Middle East that is facing a severe water crisis. The country is one of the most water-scarce nations in the world, and climate change is exacerbating the problem by increasing drought and decreasing rainfall. This is making it extremely difficult for Yemen to adapt to the impacts of climate change, such as water scarcity and food insecurity.

Somalia is a country in East Africa that is facing a severe drought and food crisis. The country has experienced a series of droughts in recent years, which are linked to climate change. This is making it extremely difficult for Somalia to adapt to the impacts of climate change, such as water scarcity and food insecurity.

Haiti is a country in the Caribbean that is facing a severe deforestation crisis. The country has lost much of its forest cover due to deforestation, which is linked to climate change. This is making it extremely difficult for Haiti to adapt to the impacts of climate change, such as flooding and landslides.

gewo-intl.org/the-threat-to-earth/

Kiribati is a small island nation in the Pacific Ocean that is facing a severe freshwater crisis. The country has limited freshwater resources and is highly vulnerable to saltwater intrusion due to rising sea levels, which are linked to climate change. This is making it extremely difficult for Kiribati to adapt to the impacts of climate change, such as water scarcity and flooding.

These examples demonstrate how poor nations with limited natural resources face additional challenges in adapting to the impacts of climate change, which can exacerbate poverty and inequality. It is important for the global community to provide support and resources to help these nations adapt to the impacts of climate change and build resilience for the future.

Dependence on fossil fuels: Many poor nations rely on fossil fuels to fuel their economic growth and development, as they may not have the resources to invest in renewable energy alternatives. This dependence on fossil fuels contributes to global emissions and makes it difficult for poor nations to reduce their carbon footprint.

There are several examples of poor nations that depend heavily on fossil fuels to drive their economy. Some of these countries include:

Nigeria: Nigeria is a major oil-producing country, with oil accounting for around 80% of its export earnings and over 90% of its foreign exchange earnings. The country heavily relies on oil revenue to fund its budget and support its economy, which makes it difficult for Nigeria to support carbon dioxide reduction and net-zero initiatives.

Venezuela: Venezuela is another country that heavily relies on its oil reserves to support its economy. The country's economy is largely dependent on oil exports, which account for around 95% of its export earnings. The political and economic turmoil in Venezuela has made it difficult for the country to support carbon dioxide reduction initiatives.

Bangladesh: Although Bangladesh is not a major producer of fossil fuels, it is heavily dependent on them for its energy needs. Around 60% of the country's energy comes from natural gas, while coal and oil also contribute significantly to the country's energy mix. The country's rapid industrialization and economic growth have made it difficult for Bangladesh to support carbon dioxide reduction initiatives.

Mozambique: Mozambique is a developing country in Southern Africa that is rich in natural resources, including coal and natural gas. The country's economy is heavily reliant on the extraction and export of these resources, which makes it difficult for Mozambique to support carbon dioxide reduction initiatives.

These poor nations face significant challenges in reducing their carbon footprint and transitioning to renewable energy sources. They often lack the financial resources and technical expertise needed to support large-scale renewable energy projects. Additionally, the economic benefits that fossil fuels provide are often difficult to replace, making it hard for these countries to fully commit to carbon dioxide reduction and net-zero initiatives.

Vulnerability to climate change impacts: Poor nations are often located in regions that are more vulnerable to the impacts of climate change, such as increased frequency of extreme weather events like droughts, floods, and hurricanes. These events can cause significant damage to infrastructure, livelihoods, and the economy, making it even more difficult for poor nations to invest in climate mitigation and adaptation measures.

Bangladesh is a densely populated country in South Asia that is highly vulnerable to the impacts of climate change, including flooding, sea level rise, and cyclones. The country is prone to frequent and devastating floods, which can damage crops, infrastructure, and homes. In 1998, Bangladesh experienced severe flooding that affected over 30 million people and caused an estimated $2.2 billion in damages. The floods were exacerbated by rising sea levels and heavy rainfall, which are both linked to climate change.

Malawi is a landlocked country in Southern Africa that is highly vulnerable to the impacts of climate change, including droughts and floods. The country is prone to frequent droughts, which can lead to crop failures and food insecurity. In 2015–2016, Malawi experienced a severe drought that affected over 6.5 million people and caused an estimated $1.1 billion in damages. The drought was linked to the El Niño weather pattern, which is made worse by climate change.

Haiti is a small island nation in the Caribbean that is highly vulnerable to the impacts of climate change, including hurricanes, flooding, and landslides. The country is prone to frequent and devastating hurricanes, which can damage infrastructure, homes, and crops. In 2016, Hurricane Matthew struck Haiti, causing widespread damage and destruction. The hurricane affected over 2 million people and caused an estimated $2.8 billion in damages. The intensity of the hurricane was linked to warming ocean temperatures, which are a result of climate change.

Kiribati is a small island nation in the Pacific Ocean that is highly vulnerable to the impacts of climate change, including sea level rise, flooding, and coastal erosion. The country is at risk of being completely submerged by rising sea levels, which could displace its entire population. Kiribati is experiencing increasing coastal erosion and flooding due to rising sea levels, which are linked to climate change. In 2015, the country declared a state of emergency due to a severe drought that affected over 50,000 people and caused widespread crop failure and food insecurity.

These examples demonstrate the devastating impacts that climate change can have on poor nations, including loss of life, displacement, and economic damages. It is important to note that these impacts are expected to worsen in the coming years as the planet continues to warm. These poor nations often lack the resources and technology needed to effectively address climate change and protect their citizens from the impacts of extreme weather events. As a result, they are often left to bear the brunt of the impacts of climate change, which can exacerbate poverty, inequality, and social instability.

Limited political power: Poor nations may not have the political power or influence necessary to negotiate favorable climate policies at international forums, leading to a lack of investment in climate mitigation and adaptation measures.

Haiti is one of the poorest countries in the world and is highly vulnerable to the impacts of climate change, including hurricanes, flooding, and sea level rise. However, the country has limited political power and influence at international climate negotiations, making it difficult for it to access the funds and resources needed to implement climate mitigation and adaptation measures.

Chad is a landlocked country in Central Africa that is highly vulnerable to the impacts of climate change, including droughts and desertification. However, the country has limited political power and influence at international climate negotiations, making it difficult for it to access the funds and resources needed to implement climate mitigation and adaptation measures.

Cambodia is a developing country in Southeast Asia that is highly vulnerable to the impacts of climate change, including flooding and sea level rise. However, the country has limited political power and influence at international climate negotiations, making it difficult for it to access the funds and resources needed to implement climate mitigation and adaptation measures.

Malawi is a landlocked country in Southern Africa that is highly vulnerable to the impacts of climate change, including droughts and flooding. However, the country has limited political power and influence at international climate negotiations, making it difficult for it to access the funds and resources needed to implement climate mitigation and adaptation measures.

These poor nations often lack the resources and technology needed to effectively address climate change, and their limited political power and influence can make it difficult for them to access the funds and resources needed to implement climate mitigation and adaptation measures. As a result, these countries are often left to bear the brunt of the impacts of climate change, which can exacerbate poverty, inequality, and social instability.

Paris Agreement and Rich Nations Contributions Towards the Fight Against Climate Change

The Paris Agreement

The Paris Agreement is a legally binding international agreement adopted in December 2015 by the United Nations Framework Convention on Climate Change (UNFCCC). The agreement aims to limit global warming to well below 2 degrees Celsius above pre-industrial levels, and to pursue efforts to limit warming to 1.5 degrees Celsius. The Paris Agreement calls for all nations to take action to reduce greenhouse gas emissions and adapt to the impacts of climate change. However, rich nations, which are primarily responsible for the historical emissions that have caused the current climate crisis, have a particular responsibility to take more significant action to address the issue. In this book, I examined the Paris Agreement and the contributions of rich nations towards the fight against climate change, and I have evaluated the effectiveness of their efforts.

Paris Agreement: Key Features

The Paris Agreement has several key features that are essential for addressing the challenges posed by climate change. Firstly, the agreement requires all nations to submit national climate action plans known as Nationally Determined Contributions (NDCs). These plans outline the actions that each country will take to reduce their greenhouse gas emissions and adapt to the impacts of climate change. The NDCs are reviewed and updated every five years to ensure that they remain ambitious and up to date.

Secondly, the Paris Agreement establishes a long-term goal of net-zero greenhouse gas emissions in the second half of the century. This means that all nations will need to reduce their greenhouse gas emissions to zero or offset their emissions through carbon removal technologies such as afforestation, reforestation, and carbon capture and storage.

Thirdly, the Paris Agreement includes a financial mechanism to support developing nations in their efforts to reduce greenhouse gas emissions and adapt to the impacts of climate change. The financial mechanism includes a goal of mobilizing $100 billion per year in climate finance from developed nations to developing nations by 2020, and a commitment to increase financial flows to developing nations beyond 2020.

Rich Nations Contributions Towards the Fight against Climate Change

Rich nations, which are primarily responsible for the historical emissions that have caused the current climate crisis, have a particular responsibility to take significant action to address the issue. The contributions of rich nations towards the fight against climate change can be broadly categorized into two areas: reducing their greenhouse gas emissions and providing financial support to developing nations.

Reducing Greenhouse Gas Emissions

Rich nations have made significant progress in reducing their greenhouse gas emissions over the past decade. The European Union, for example, has committed to reducing its emissions by at least 55% by 2030, compared to 1990 levels. The United States, under the Obama administration, pledged to reduce its emissions by 26–28% below 2005 levels by 2025. Other nations, such as Japan and Canada, have also made commitments to reduce their emissions.

However, some critics argue that these commitments are not ambitious enough to limit warming to well below 2 degrees Celsius or pursue efforts to limit warming to 1.5 degrees Celsius, as required by the Paris Agreement. For example, the Intergovernmental Panel on Climate Change (IPCC) has called for global emissions to decline by 45% by 2030, compared to 2010 levels, and reach net-zero by 2050 to limit warming to 1.5 degrees Celsius.

Providing Financial Support to Developing Nations

Rich nations have also committed to providing financial support to developing nations to help them reduce their greenhouse gas emissions and adapt to the impacts of climate change. The Paris Agreement calls for the mobilization of $100 billion per year in climate finance, from developed nations to developing nations, by 2020, and a commitment to increase financial flows to developing nations beyond 2020.

However, the actual amount of climate finance provided by rich nations has fallen short of the $100 billion per year goal.

According to the Organisation for Economic Co-operation and Development (OECD), climate finance from developed nations reached $79.6 billion in 2019, which is an increase from previous years but still falls short of the $100 billion per year goal. Additionally, a significant portion of climate finance provided by rich nations is in the form of loans rather than grants, which can be challenging for developing nations to repay.

Furthermore, some critics argue that rich nations have not done enough to address the issue of loss and damage caused by climate change in developing nations. Loss and damage refer to the impacts of climate change that cannot be addressed through adaptation measures, such as the loss of biodiversity, cultural heritage, and human lives. The Paris Agreement established the Warsaw International Mechanism for Loss and Damage, which aims to address this issue. However, critics argue that rich nations have not provided sufficient financial support to address loss and damage in developing nations.

Effectiveness of Rich Nations' Contributions

The effectiveness of rich nations' contributions towards the fight against climate change can be evaluated based on several factors, including the scale of their emissions reductions, the level of financial support provided to developing nations, and the overall impact of their actions on global greenhouse gas emissions.

In terms of emissions reductions, while rich nations have made progress in reducing their greenhouse gas emissions, their commitments are not ambitious enough to limit warming to well below 2 degrees Celsius or pursue efforts to limit warming to 1.5 degrees Celsius, as required by the Paris Agreement. Additionally, some rich nations, such as the United States under the Trump administration, have rolled back climate policies and increased their greenhouse gas emissions, which undermines global efforts to address climate change.

In terms of financial support provided to developing nations, rich nations have not met the $100 billion per year goal established by the Paris Agreement. While the amount of climate finance provided has increased in recent years, it falls short of the amount required to support developing nations in their efforts to reduce greenhouse gas emissions and adapt to the impacts of climate change. Additionally, a significant portion of climate finance provided is in the form of loans rather than grants, which can be challenging for developing nations to repay.

gewo-intl.org/the-threat-to-earth/

Overall, the impact of rich nations' contributions on global greenhouse gas emissions is limited. While their emissions reductions and financial support provided to developing nations are important, they are not sufficient to address the scale of the climate crisis. To limit warming to well below 2 degrees Celsius or pursue efforts to limit warming to 1.5 degrees Celsius, global emissions will need to decline much more rapidly, and all nations will need to significantly increase their efforts to address the issue.

In conclusion, the Paris Agreement is an essential international agreement that aims to limit global warming to well below 2 degrees Celsius and pursue efforts to limit warming to 1.5 degrees Celsius. Rich nations, which are primarily responsible for the historical emissions that have caused the current climate crisis, have a particular responsibility to take significant action to address the issue. While rich nations have made progress in reducing their greenhouse gas emissions and providing financial support to developing nations, their efforts fall short of what is required to address the scale of the climate crisis. To limit warming to well below 2 degrees Celsius or pursue efforts to limit warming to 1.5 degrees Celsius, all nations will need to significantly increase their efforts to address the issue.

Enhancing the Contributions of Rich Nations to the Fight Against Climate Change

Here are the steps that rich nations can take to enhance their contributions to the fight against climate change:

Increase their emissions reduction targets: Rich nations should increase their emissions reduction targets to align with the goal of limiting global warming to well below 2 degrees Celsius or pursuing efforts to limit warming to 1.5 degrees Celsius. This will require significant investments in renewable energy, energy efficiency, and other low-carbon technologies.

Provide additional financial support to developing nations: Rich nations should increase their financial support to developing nations to help them reduce their greenhouse gas emissions and adapt to the impacts of climate change. This support should be provided in the form of grants rather than loans to ensure that it is accessible to developing nations.

Address loss and damage: Rich nations should provide additional financial support to address the impacts of loss and damage caused by climate change in developing nations. This support should be provided in addition to climate finance provided to support emissions reductions and adaptation measures.

Support technology transfer: Rich nations should support the transfer of low-carbon technologies to developing nations to help them reduce their greenhouse gas emissions and enhance their climate resilience. This support can include funding for research and development, as well as technology transfer agreements between developed and developing nations.

Take leadership roles: Rich nations should take leadership roles in international climate negotiations and encourage other nations to take significant action to address the issue. This can include supporting the development of ambitious emissions reduction targets and advocating for strong climate policies and regulations.

Supporting forest conservation and restoration: Forests play a critical role in mitigating climate change by absorbing carbon dioxide from the atmosphere. However, deforestation and forest degradation are significant contributors to greenhouse gas emissions. Rich nations have a responsibility to support forest conservation and restoration efforts in developing nations to help mitigate the impacts of climate change.
Rich nations can provide financial support to developing nations to support sustainable forest management practices and the implementation of REDD+ (Reducing Emissions from Deforestation and Forest Degradation) programs. These programs can incentivize forest

conservation and restoration by providing financial compensation for avoided deforestation and emissions reductions.

Promoting sustainable agriculture: Agriculture is a significant contributor to greenhouse gas emissions, primarily through livestock production and fertilizer use. Rich nations have a responsibility to promote sustainable agriculture practices that reduce greenhouse gas emissions and enhance climate resilience.

Rich nations can support the development of sustainable agriculture practices by providing financial support for research and development, as well as implementing policies and regulations to incentivize sustainable land use practices. This can include promoting the adoption of climate-smart agriculture techniques, such as agroforestry and conservation agriculture.

Encouraging low-carbon transportation: Transportation is a significant contributor to greenhouse gas emissions, particularly in developed nations. Rich nations have a responsibility to encourage low-carbon transportation options to reduce emissions and enhance climate resilience.

Rich nations can support the adoption of low-carbon transportation options by investing in public transportation infrastructure, promoting the adoption of electric vehicles, and implementing policies and regulations to incentivize the transition to low-carbon transportation. This can include implementing fuel efficiency standards, promoting the use of alternative fuels, and implementing carbon pricing mechanisms.

Promoting energy efficiency: Energy efficiency is a critical component of reducing greenhouse gas emissions and enhancing climate resilience. Rich nations have a responsibility to promote energy efficiency practices in their own countries and support energy efficiency initiatives in developing nations.

Rich nations can promote energy efficiency by implementing policies and regulations to incentivize the adoption of energy-efficient technologies, such as building codes and energy efficiency standards. Rich nations can also support energy efficiency initiatives in developing nations by providing financial support for research and development, as well as implementing policies and regulations to promote energy efficiency practices.

Addressing climate migration: The impacts of climate change can result in forced migration as individuals and communities are displaced by rising sea levels, natural disasters, and other impacts. Rich nations have a responsibility to address the issue of climate migration and support the rights of climate refugees.

Rich nations can address climate migration by implementing policies and regulations to support climate refugees, such as providing legal pathways for migration and supporting the resettlement of climate refugees. Rich nations can also provide financial support to developing nations to address the impacts of climate change and reduce the need for climate migration.

In addition to the efforts of rich nations, it is also important to note the role of civil society and non-state actors in the implementation of the Paris Agreement. Civil society organizations, including NGOs and grassroots movements, play a critical role in advocating for climate action and holding governments and corporations accountable for their commitments to the Paris Agreement. Non-state actors, including businesses, cities, and states, also play a critical role in implementing the agreement and reducing emissions.

For example, many cities and states in the United States have committed to meeting the goals of the Paris Agreement, even after the country withdrew from the agreement. This includes cities like New York and Los Angeles, which have committed to reducing their emissions by 80% by

2050. Similarly, companies like Google and Apple have committed to using 100% renewable energy in their operations.

However, it is important to note that the efforts of non-state actors cannot replace the need for strong action from governments and international institutions. While the efforts of non-state actors are important, they cannot substitute the need for comprehensive and coordinated action at the national and international levels.

The Paris Agreement represents a critical step forward in the global effort to address climate change. However, much more needs to be done to achieve the goals of the agreement and limit the impacts of climate change on poor nations and vulnerable communities. Rich nations have a significant responsibility to support poor nations in their efforts to address climate change and prioritize the needs and concerns of these communities. Civil society organizations and non-state actors also play a critical role in implementing the agreement and advocating for climate action. By working together, we can create a more sustainable and equitable future for all.

Looking ahead, the success of the Paris Agreement will depend on the ability of nations to accelerate the transition to a low-carbon and climate-resilient economy. This will require the development and deployment of new technologies, investments in clean energy and sustainable infrastructure, and the adoption of policies and regulations that support the transition to a low-carbon economy.

At the same time, it is important to recognize that the impacts of climate change will continue to be felt by poor nations and vulnerable communities for decades to come, even as efforts to mitigate and adapt to climate change continue. As such, it is critical that rich nations continue to prioritize the needs and concerns of these communities and provide them with the support and resources necessary to build resilience and adapt to the impacts of climate change.

In this regard, the concept of "loss and damage" has emerged as a critical issue in the international climate negotiations. Loss and damage refer to the impacts of climate change that cannot be addressed through mitigation or adaptation measures, such as the loss of lives, livelihoods, and cultural heritage. The Paris Agreement recognizes the importance of addressing loss and damage, and established the Warsaw International Mechanism for Loss and Damage to address this issue.

However, much more needs to be done to address the needs and concerns of poor nations and vulnerable communities, particularly as the impacts of climate change continue to intensify. This will require the sustained commitment of rich nations and international institutions to support the efforts of these communities to adapt to the impacts of climate change and build resilience.

The Paris Agreement represents a critical step forward in the global effort to address climate change, but much more needs to be done to achieve the goals of the agreement and address the needs and concerns of poor nations and vulnerable communities. Rich nations have a critical responsibility to support these communities and provide them with the resources and support necessary to build resilience and adapt to the impacts of climate change. By working together, we can create a more sustainable and equitable future for all.

In addition to the ongoing efforts to address climate change, it is also important to recognize the interconnected nature of the challenges facing the global community. Climate change is not just an environmental issue, but a social, economic, and political one as well, with implications for issues such as public health, human rights, and social justice.

As such, addressing climate change will require a holistic approach that recognizes the need for systemic change and transformative action. This will require not only the deployment of new technologies and the implementation of policy measures, but also a fundamental shift in the way we think about and approach development.

Ultimately, the success of the Paris Agreement and the global fight against climate change will depend on the ability of nations and communities to work together towards a common goal. This will require a shared sense of purpose and a commitment to collaboration and cooperation, as well as the recognition of the need to prioritize the needs and concerns of the most vulnerable members of society.

In the years to come, it is essential that we continue to build on the progress made under the Paris Agreement and work towards a more sustainable, just, and equitable future for all. By doing so, we can create a world that is resilient to the impacts of climate change, promotes human well-being and social justice, and ensures the long-term viability of our planet.

Key Challenges Facing the Global Community in the Fight Against Climate Change

One of the key challenges facing the global community in the fight against climate change is the need to address the social and economic dimensions of the issue. While climate change is often framed as an environmental problem, its impacts are felt across a range of sectors, including health, agriculture, energy, and water. As such, addressing climate change will require a holistic approach that recognizes the need for systemic change and transformative action.

This means taking a comprehensive view of development and recognizing the importance of sustainable, equitable, and inclusive growth. It also means addressing the social and economic factors that contribute to climate vulnerability, such as poverty, inequality, and marginalization. In many cases, these factors are intertwined, with poverty and social exclusion exacerbating the impacts of climate change and limiting the ability of communities to adapt to its effects.

To address these challenges, it will be essential to promote sustainable and inclusive economic growth, while also prioritizing investments in social protection and disaster risk reduction. This will require a coordinated approach that brings together governments, civil society organizations, and the private sector, as well as a commitment to scaling up investment in clean energy, sustainable agriculture, and resilient infrastructure.

At the same time, it is also important to recognize the need for systemic change in the way we approach development. This will require a fundamental shift away from business-as-usual approaches that prioritize short-term economic gain over long-term sustainability. It will also require a recognition of the importance of environmental protection, as well as the need to address social and economic inequalities and promote social justice.

As we move forward in the fight against climate change, it will be essential to prioritize the needs and concerns of the most vulnerable members of society, including low-income communities, women, children, and marginalized groups. This will require a commitment to inclusive decision-making processes and a recognition of the importance of social participation and empowerment.

Ultimately, the success of the Paris Agreement and the global fight against climate change will depend on the ability of nations and communities to work together towards a common goal. This will require a shared sense of purpose and a commitment to collaboration and cooperation, as well as the recognition of the need to prioritize the needs and concerns of the most vulnerable members of society.

Chapter 6

Dooms Day Is Coming

Endangered Cities That Will Disappear in the Next 100 Years Because of Climate Change

Miami, USA

Miami, USA, is one of the most at-risk cities in the world when it comes to rising sea levels. The city sits on a limestone foundation that is highly porous and vulnerable to erosion, and much of Miami is located at or below sea level, which makes it highly susceptible to flooding. A report by the National Oceanic and Atmospheric Administration (NOAA) predicts that sea levels in Miami could rise by as much as 6 feet by the end of the century, which would make the city virtually uninhabitable.

The city has already experienced a number of extreme weather events in recent years, including Hurricane Irma in 2017, which caused extensive damage to buildings and infrastructure and left much of the city without power for several days. Miami has also experienced more frequent and severe flooding in recent years, including during high tides and rainstorms. As a result, the city is already experiencing the effects of sea level rise, as mentioned earlier.

Miami is home to over 5 million people and is a popular tourist destination. However, a 3-meter sea level rise would inundate much of the city, including its downtown area and major airports, which could be devastating to the city's economy and public health.

Miami's vulnerability to sea level rise is due to its low elevation and the porous limestone geology of the region. According to a study by the Union of Concerned Scientists, Miami is one of the cities in the United States most at risk of chronic flooding due to sea level rise by 2100. In addition, the Miami-Dade Sea Level Rise Task Force predicts that the city could experience sea level rise of up to 2 feet by 2060, which could lead to flooding of streets, homes, and businesses.

Miami is already experiencing the effects of rising sea levels, with sea levels in the city having risen by 3–4 inches since the 1990s, according to a report by the Southeast Florida Regional Climate Change Compact. The city has experienced more frequent and severe flooding in recent years, which poses a serious threat to public health and safety, in addition to being a nuisance.

In conclusion, Miami is a major city at risk of disappearing due to the effects of climate change, particularly rising sea levels. The city is already experiencing the effects of sea level rise, including flooding during high tides, and predictions suggest that the situation is only going to get worse. It is crucial that measures are taken to mitigate the effects of climate change and protect the city and its residents.

Venice, Italy

Venice, Italy, is a historic city built on a group of islands in the Venetian Lagoon, home to over 260,000 people. However, the city is facing multiple challenges due to the effects of climate change. Venice is already sinking at a rate of about 1–2 millimeters per year due to natural subsidence, and rising sea levels are exacerbating the problem. The city is vulnerable to flooding due to extreme weather events like storm surges. In fact, in November 2019, Venice experienced the highest tide in 50 years, which caused extensive flooding and damage to the city's historic buildings and infrastructure.

The flooding reached a level of 1.87 meters above sea level, the second highest on record. This event highlighted the vulnerability of Venice to the effects of climate change, including rising sea levels and extreme weather events. The city has implemented measures to protect against flooding, but some experts believe that these measures may not be enough to prevent the city from disappearing entirely.

Moreover, a report by the City of Venice has predicted that sea level rise will cause significant flooding and erosion in the city over the next century, with the iconic St. Mark's Square flooding more frequently. The city's location on the coast, coupled with the fact that much of the city is already below sea level, makes it even more vulnerable to sea level rise. A 3-meter sea level rise could significantly increase the risk of flooding in the city and cause significant damage to its historic buildings.

In summary, Venice's unique location and historical significance are under threat from the effects of climate change. The city is already experiencing the impacts of rising sea levels and extreme weather events, and unless significant action is taken, Venice may face a bleak future.

Dhaka

Dhaka, the capital city of Bangladesh, is one of the fastest-growing cities in the world and home to over 21 million people. However, the city is also highly vulnerable to the impacts of climate change. Dhaka is located in a low-lying area and is prone to flooding during the monsoon season, which can cause extensive damage to buildings and infrastructure, as well as displacement of residents. In 2020, the city experienced some of the worst flooding in decades, affecting over 2 million people and leading to food shortages in some areas.

Moreover, Dhaka is also vulnerable to sea level rise due to its location on the coast of the Bay of Bengal and its low-lying areas. According to a study by Climate Central, Dhaka is one of the cities at the highest risk of being impacted by sea level rise by 2100, with over 40% of the city's land area at risk of being flooded. A 3-meter sea level rise would inundate much of the city's coastal areas, including its major port facilities and airport.

Dhaka is already experiencing the effects of climate change, including more frequent and severe flooding during the monsoon season. In addition, the city's groundwater is being depleted due to rising sea levels and saltwater intrusion. This is leading to water shortages and could threaten the city's ability to support its growing population. Some experts believe that Dhaka could become uninhabitable within the next few decades if climate change is not addressed.

In summary, Dhaka is a highly vulnerable city facing the impacts of climate change, including more frequent and severe flooding, water shortages, and the risk of becoming uninhabitable in the future. Urgent action is needed to address these issues and protect the city and its residents.

Lagos, Nigeria

Lagos, Nigeria, is one of the largest cities in the world, home to over 24 million people. However, the city's location on the coast makes it highly vulnerable to the effects of climate change. Due to rising sea levels, Lagos has already experienced a number of extreme weather events such as erosion and flooding. In fact, the city is losing about 2 meters of land per year to erosion, according to a report by the World Bank. The city's rapid development has also led to the destruction of natural habitats, which could exacerbate the effects of climate change in the future.

gewo-intl.org/the-threat-to-earth/

In 2017, Lagos experienced a major flood that caused extensive damage to buildings and infrastructure and left many residents homeless. This flooding was caused by heavy rainfall and poor drainage systems, which are being worsened by the effects of climate change. A report by the African Development Bank has also identified Lagos as being at high risk of flooding due to its low elevation and the impact of climate change.

A most recent flood hazard in Lagos, Nigeria, as at the time I wrote this book, was in 2021. The floods paralyzed economic activity, at an estimated cost of around $4 billion per year, and resulted in an estimated $17 billion damage. In 2019, a total of 277,555 people were affected by floods and 158 were killed in Lagos. More than 600 people have died in the worst flooding the West African nation has seen in a decade. Some 1.3 million people have been displaced, and more than 200,000 homes have been destroyed.

The effects of sea level rise are a particular concern for Lagos, with a 3-meter sea level rise posing a significant threat to much of the city's low-lying areas, including major airports and port facilities. Given its current vulnerability and the expected impacts of climate change, it is crucial for Lagos to take action to mitigate and adapt to these threats, such as by implementing better drainage systems and protecting natural habitats.

Jakarta, Indonesia

Jakarta, Indonesia, is facing multiple environmental challenges. The city is one of the fastest-sinking cities in the world, with some areas sinking as much as 25 centimeters per year. The sinking is caused by natural subsidence, groundwater extraction, and the weight of buildings and infrastructure, which are causing extensive damage to the city's infrastructure, including roads, buildings, and pipelines.

In addition to sinking, Jakarta is also prone to flooding due to its location on the coast and the many rivers that run through it. In 2020, the city experienced some of the worst flooding in years, with over 300,000 people affected and at least 60 dead. The flooding was caused by heavy rainfall and poor drainage systems, which are being exacerbated by the effects of climate change.

Furthermore, Jakarta is one of the most polluted cities in the world, which could worsen the impact of climate change on the city's inhabitants. The city is home to over 10 million people and is located on the coast of the Java Sea, making it vulnerable to sea level rise. A 3-meter sea level rise would inundate much of the city's low-lying areas, including its airport and major port facilities. A report by the Jakarta Post indicates that the city could experience sea level rise of up to 1.5 meters by 2100, which could lead to flooding and displacement of millions of people.

Shanghai, China

Shanghai, China, is a major city located on the coast of the East China Sea and is home to over 23 million people. The city is vulnerable to sea level rise due to its low-lying areas and location on the Yangtze River Delta. In addition, Shanghai is sinking at an alarming rate due to groundwater extraction, which exacerbates the effects of sea level rise.

A 3-meter sea level rise would have disastrous consequences for Shanghai, as it would inundate much of the city, including its financial district and major airports. The city's port facilities and low-lying areas would also be at risk of flooding. Shanghai is taking steps to address these challenges, including building protective sea walls and improving its drainage systems.

Overall, the combination of sinking land and sea level rise poses a significant threat to Shanghai's infrastructure, economy, and millions of residents. Addressing these challenges will require a coordinated effort from the government, businesses, and the public to mitigate the effects of climate change and protect the city's future.

Mumbai, India

Mumbai, India, is one of the largest cities in the world, with a population of over 18 million people. The city is located on the coast of the Arabian Sea, which makes it vulnerable to sea level rise. Mumbai is already prone to flooding during the monsoon season, and a 3-meter sea level rise would inundate much of the city's low-lying areas, including its airport and major port facilities.

According to a study by the Indian Institute of Technology Bombay, Mumbai is at high risk of flooding due to sea level rise, particularly during extreme weather events such as cyclones. The study also found that the city's vulnerability to flooding is compounded by its rapidly growing population and urbanization. In addition to sea level rise, Mumbai is also facing the threat of sinking due to groundwater extraction.

Despite these risks, Mumbai has been slow to implement measures to mitigate the effects of climate change. The city has recently begun construction of a coastal road, which has faced criticism from environmental activists who argue that it could exacerbate the city's vulnerability to sea level rise and flooding.

As one of the largest cities in the world, Mumbai's vulnerability to sea level rise and flooding poses a significant threat to its residents and economy. It is important for the city and its leaders to prioritize climate action and implement measures to adapt to the effects of climate change.

New York City, USA

New York City, USA, is one of the most populous cities in the world with over 8 million people, and it is also vulnerable to sea level rise due to its location on the coast. The city has already experienced the effects of sea level rise, with Hurricane Sandy causing widespread damage and flooding in 2012.

A 2019 report by the New York City Panel on Climate Change warns that the city could experience sea level rise of up to 6 feet by 2100. This could lead to regular flooding, damage to infrastructure, and displacement of millions of people. The report also notes that the city's low-lying areas, including its financial district and major airports, are particularly vulnerable to sea level rise.

The Hudson River, which flows through the heart of New York City, is also at risk of flooding due to sea level rise. In addition to the risk of coastal flooding, the city's aging infrastructure is already struggling to cope with current levels of rainfall and is at risk of being overwhelmed by more frequent and intense storms caused by climate change.

To combat these risks, New York City has implemented various measures such as a $20 billion plan to build a system of flood barriers and levees, known as the "Big U," to protect lower Manhattan from flooding. The city has also set a goal of reducing its greenhouse gas emissions by 80% by 2050, through initiatives such as promoting energy efficiency and expanding the use of renewable energy sources.

Despite these efforts, the effects of sea level rise are expected to be felt in New York City for decades to come and will require ongoing adaptation and investment in infrastructure to mitigate the damage.

Bangkok

Bangkok, the capital city of Thailand, is home to over 10 million people and is located on the coast of the Gulf of Thailand. The city is vulnerable to sea level rise due to its low-lying areas and its location on the Chao Phraya River delta. A 3-meter sea level rise would inundate much of the city's coastal areas, including its major port facilities and airport, which are crucial to the city's economy.

According to a study by Climate Central, Bangkok is one of the cities at highest risk of being impacted by sea level rise by 2100. The study found that over 40% of the city's land area is at risk of being flooded, and millions of people could be displaced if measures to mitigate the effects of climate change are not taken.

The impacts of sea level rise in Bangkok are already being felt, as the city has experienced more frequent and severe flooding in recent years, particularly during the monsoon season. In addition to the threat of flooding, sea level rise could also worsen coastal erosion, saltwater intrusion into freshwater sources, and damage to infrastructure.

In response to these risks, the Thai government has launched several initiatives aimed at reducing greenhouse gas emissions and adapting to the impacts of climate change, including the development of a climate change master plan and investments in flood prevention and water management infrastructure. However, there is still much work to be done to ensure the long-term resilience of the city and its people in the face of sea level rise and other climate-related challenges.

Rio de Janeiro, Brazil

Rio de Janeiro is a major coastal city located in Brazil and is home to over 6 million people. Due to its location on the coast, it is vulnerable to the impacts of sea level rise. According to a study by the World Bank, a 1-

meter sea level rise would put over 128,000 people and $16 billion worth of property at risk of flooding in Rio de Janeiro.

Rio de Janeiro is not only threatened by sea level rise, but also by extreme weather events such as floods and landslides. These events are expected to increase in frequency and intensity due to climate change. A study by the University of Bristol found that in the future, Rio de Janeiro will be at greater risk of floods and landslides as a result of climate change.

In addition to the risks posed to human life and property, sea level rise also poses a significant threat to Rio de Janeiro's unique coastal ecosystems, such as mangroves and coral reefs. These ecosystems play a crucial role in protecting against coastal erosion, buffering the impacts of storms, and providing habitat for marine species. The loss of these ecosystems due to sea level rise could have far-reaching impacts on the city's environment and economy.

Alexandria, Egypt

Alexandria, Egypt, is a historic port city located on the coast of the Mediterranean Sea and home to over 5 million people. Due to its low-lying areas and location on the coast, the city is vulnerable to sea level rise. A 3-meter sea level rise would inundate much of the city's coastal areas, including its historic buildings and major transportation hubs, as well as its major port facilities.

Furthermore, Alexandria is already experiencing the impacts of sea level rise, with flooding becoming increasingly common in recent years. A study by the Arab Academy for Science, Technology and Maritime Transport identified Alexandria as one of the cities most at risk of being inundated by sea level rise in the Mediterranean region. The study noted that Alexandria's coastal areas have already experienced significant land loss due to sea level rise, with estimates suggesting that the city has lost around 20% of its land area in the last century.

Given the city's vulnerability to sea level rise, experts are calling for action to protect Alexandria and its people from the impacts of climate change. Proposed solutions include building seawalls, implementing flood management systems, and relocating people and infrastructure away from high-risk areas.

Ho Chi Minh City, Vietnam

Ho Chi Minh City, also known as Saigon, is a major coastal city in Vietnam with a population of over 8 million. It is located on the coast of the South China Sea, which makes it susceptible to the adverse effects of sea level rise. Rising sea levels are likely to inundate many of the city's low-lying areas, including crucial transportation infrastructure such as airports and port facilities. If the sea level were to rise by 3 meters, it would have a significant impact on the city.

There is already evidence of sea level rise in the region. A 2019 report by Climate Central found that Ho Chi Minh City is among the 10 cities in the world most at risk of chronic flooding from sea level rise by 2050. The report found that nearly 40% of the city would be at risk of flooding, affecting over 6 million people.

According to the World Bank, Vietnam is one of the most vulnerable countries to climate change in the world. The country is already experiencing more frequent and severe natural disasters, including flooding and typhoons, which have caused significant damage to infrastructure and loss of life.

In conclusion, Ho Chi Minh City, with its location on the coast of the South China Sea and low-lying areas, is highly vulnerable to sea level rise. The effects of rising sea levels would have significant impacts on the city's transportation infrastructure and the livelihoods of its 8 million inhabitants. The need for urgent action to mitigate and adapt to the impacts of climate change in Vietnam cannot be overstated.

Karachi, Pakistan

Karachi is a major city in Pakistan, located on the coast of the Arabian Sea. It is home to over 15 million people, making it one of the most populous cities in the world. However, the city's location on the coast also makes it vulnerable to the effects of climate change, particularly sea level rise.

If sea levels were to rise by 3 meters, much of Karachi's coastal areas would be inundated, including its major port facilities. This could have severe implications for the city's economy, as the port of Karachi is one of the busiest in the region, handling over 60% of Pakistan's cargo.

According to a report by the Asian Development Bank, Karachi is at high risk of flooding due to climate change. The report predicts that rising sea levels could lead to frequent and severe flooding, particularly during monsoon season.

This is not the only climate-related challenge facing Karachi. The city also experiences intense heatwaves, with temperatures regularly exceeding 40°C (104°F) during the summer months. These heatwaves have caused significant loss of life in the past, particularly among vulnerable populations such as the elderly and those living in informal settlements.

In conclusion, Karachi's location on the coast of the Arabian Sea makes it highly vulnerable to the impacts of climate change, particularly sea level rise and flooding. The potential loss of the city's major port facilities would have significant implications for the local and national economy. Urgent action is needed to mitigate and adapt to the effects of climate change in Karachi and other vulnerable cities in the region.

gewo-intl.org/the-threat-to-earth/

Alexandria, Virginia, USA

Alexandria is a small historic city located on the coast of the Potomac River in Virginia, USA. It is home to over 150,000 people and is already experiencing the effects of sea level rise. The city's low-lying areas, including its historic buildings and major transportation hubs, are particularly vulnerable to flooding.

A 3-meter sea level rise would have a significant impact on Alexandria, increasing the risk of flooding and causing damage to its infrastructure. The city's vulnerability to sea level rise is further confirmed by a report from the National Oceanic and Atmospheric Administration, which identifies Alexandria as one of the cities in the United States most at risk of sea level rise by 2100.

In addition to sea level rise, Alexandria is also vulnerable to other climate-related challenges, including extreme weather events such as hurricanes and storm surges. Hurricane Isabel, which hit the city in 2003, caused significant damage and flooding, particularly in low-lying areas.

To address these challenges, the city has developed a comprehensive plan to reduce its vulnerability to the impacts of climate change. This includes measures such as elevating buildings, improving stormwater management, and developing early warning systems for floods and other extreme weather events.

In conclusion, Alexandria, Virginia, is a historic city located on the coast of the Potomac River and is vulnerable to the impacts of sea level rise and other climate-related challenges. The city's infrastructure and historic buildings are particularly at risk, and urgent action is needed to adapt to the changing climate and reduce the risk of damage and loss of life.

Alexandria Bay, New York, USA

Alexandria Bay is a small town located on the banks of the St. Lawrence River in upstate New York, USA. The town is vulnerable to the impacts of sea level rise and flooding, with water levels on the river already increasing in recent years.

According to a report by Climate Central, Alexandria Bay is one of the cities in the United States most at risk of coastal flooding by 2050. The report identifies the town as one of the top 20 most vulnerable places in the country, with over 50% of its population and infrastructure at risk of flooding due to sea level rise.

In addition to sea level rise, Alexandria Bay is also vulnerable to other climate-related hazards such as extreme weather events, including heavy rainfall and storm surges. These events can cause significant damage to the town's infrastructure and homes, as was the case during the severe flooding that occurred in 2017.

To address these challenges, the town has developed a comprehensive plan to reduce its vulnerability to the impacts of climate change. This includes measures such as improving infrastructure to withstand flooding, raising public awareness of flood risks, and developing early warning systems for extreme weather events.

In conclusion, Alexandria Bay, New York, is a small town located on the banks of the St. Lawrence River and is vulnerable to the impacts of sea level rise and other climate-related hazards. Urgent action is needed to adapt to the changing climate and reduce the risk of damage and loss of life.

Alexandria, Louisiana, USA

Alexandria, Louisiana, USA is a small city located on the coast of the Red River and is home to over 47,000 people. The city is vulnerable to the impacts of sea level rise due to its low-lying areas and location on the coast.

According to a report by the National Oceanic and Atmospheric Administration, Alexandria is one of the cities in the United States most at risk of sea level rise by 2100. The report predicts that a 3-meter sea level rise would significantly increase the risk of flooding in the city and could cause significant damage to its infrastructure.

In addition to sea level rise, Alexandria is also vulnerable to other climate-related hazards such as extreme weather events, including heavy rainfall and flooding. These events can cause significant damage to the city's infrastructure, homes, and businesses.

To address these challenges, the city has developed a comprehensive plan to reduce its vulnerability to the impacts of climate change. This includes measures such as improving infrastructure to withstand flooding, developing early warning systems for extreme weather events, and promoting sustainable land use practices.

Overall, Alexandria, Louisiana, USA, is a small city located on the coast of the Red River and is vulnerable to the impacts of sea level rise and other climate-related hazards. Urgent action is needed to adapt to the changing climate and reduce the risk of damage and loss of life.

Tokyo, Japan

Tokyo, Japan, is the capital city of Japan and home to over 13 million people. The city is located on Tokyo Bay, which makes it vulnerable to the impacts of sea level rise. A 3-meter sea level rise would inundate much of the city's low-lying areas, including its major airports and port facilities. In addition, Tokyo is at risk of coastal flooding due to its extensive network of rivers and canals.

A study conducted by Climate Central found that Tokyo is one of the cities most at risk of coastal flooding by 2050, with up to 5.5 million people at risk. The study indicates that if current trends continue, coastal flooding could affect up to 15% of the city's population by 2100. The Japanese government has recognized the threat of sea level rise and has taken steps to mitigate the impacts, including the construction of floodgates and seawalls, as well as efforts to improve drainage and water management systems. However, the city continues to face significant challenges in adapting to the impacts of climate change.

Rotterdam, Netherlands

Rotterdam is a major port city located on the delta of the Rhine and Meuse rivers and the coast of the North Sea in the Netherlands. It is home to over 650,000 people and is one of the most important economic centers in Europe. However, the city is also vulnerable to the impacts of sea level rise and flooding due to its low-lying areas and location on the coast.

A 3-meter sea level rise would pose a significant threat to the city, inundating much of its port facilities and low-lying areas. In fact, according to a report by Deltares, Rotterdam is one of the cities most vulnerable to sea level rise in Europe, with sea level projected to rise by up to 2 meters by 2100.

To address this issue, Rotterdam has implemented a series of measures to increase its resilience to the impacts of sea level rise and flooding. These measures include the construction of a 360-meter-long barrier in the Nieuwe Waterweg canal, which can be closed in the event of a storm surge, and the development of floating houses and buildings that can rise and fall with changing water levels.

Despite these efforts, Rotterdam still faces significant challenges in adapting to the impacts of sea level rise and climate change. As such, the city continues to invest in research and innovation to stay ahead of these challenges and maintain its position as a global leader in sustainability and resilience.

Osaka, Japan

Osaka, Japan, is a coastal city with a population of over 19 million people. Due to its location on the coast and low elevation, it is vulnerable to the impacts of sea level rise and flooding. A study by Climate Central suggests that Osaka is one of the cities at highest risk of coastal flooding in the world by 2100.

The risk of flooding in Osaka is a significant concern, as a 3-meter sea level rise would inundate much of the city's port facilities and low-lying areas. The effects of flooding could be devastating for the city and its inhabitants.

In addition to the physical and economic impacts of flooding, climate change and sea level rise could also have significant social impacts. Displacement of people, loss of homes and businesses, and disruption of essential services could all have serious consequences for the people of Osaka.

The threat of flooding in Osaka highlights the urgent need for action to address the causes of climate change and to prepare for its impacts. Investing in infrastructure to protect against flooding, and promoting sustainable development practices, could help to mitigate the risks to the city and its people.

Manila, Philippines

Manila, the capital city of the Philippines, is home to over 13 million people and is located on the coast of Manila Bay. Due to its low-lying areas, the city is particularly vulnerable to the impacts of sea level rise. In fact, a 3-meter sea level rise would inundate much of the city's coastal areas and major port facilities.

The threat of sea level rise and flooding in Manila is not just hypothetical; it is already happening. The city experiences regular flooding during high tides, particularly in low-lying areas. In 2020, Typhoon Vamco caused severe flooding in Manila, displacing thousands of people and causing significant damage to homes and infrastructure.

The impacts of flooding in Manila are not just physical; they can also have significant social and economic consequences. Displacement, loss of homes and businesses, and disruptions to essential services can all have long-lasting effects on the people of the city.

Given the severity of the threat of sea level rise and flooding in Manila, there is an urgent need for action to address the causes of climate change and to prepare for its impacts. Investing in infrastructure to protect against flooding, promoting sustainable development practices, and implementing early warning systems can all help to mitigate the risks to the city and its people. Additionally, there is a need for greater international cooperation and support to help vulnerable cities like Manila to adapt to the impacts of climate change.

New Orleans, USA

New Orleans is a city located in Louisiana, USA, with a population of over 390,000 people. The city is particularly vulnerable to the impacts of sea level rise due to its location on the coast and the fact that much of the city is already below sea level. In fact, New Orleans is already experiencing regular flooding and land loss due to sea level rise.

The threat of sea level rise and flooding in New Orleans is particularly concerning given the city's historic buildings and major transportation hubs. A 3-meter sea level rise would significantly increase the risk of flooding and could cause significant damage to these important structures.

The city's vulnerability to sea level rise was made tragically clear during Hurricane Katrina in 2005, which caused widespread flooding and devastation in New Orleans. The disaster highlighted the urgent need for action to address the causes of climate change and to prepare for its impacts.

In recent years, the city has taken steps to protect against future flooding, including investing in infrastructure to improve drainage and levees. However, more work is needed to ensure that the city is adequately prepared for the impacts of sea level rise and other climate-related risks.

Given the importance of New Orleans to the cultural and economic fabric of the United States, there is a need for greater investment and support to help the city adapt to the impacts of climate change. This could include funding for infrastructure improvements and incentives for sustainable development practices.

Tianjin, China

Tianjin is a major port city located on the coast of the Bohai Sea in China. With a population of over 15 million people, the city is a vital economic hub for the region. However, Tianjin is also highly vulnerable to the impacts of sea level rise due to its low-lying areas and location on the coast.

A 3-meter sea level rise would have devastating consequences for the city, inundating much of its port facilities and low-lying areas. This could have significant impacts on the city's economy, as well as the well-being of its residents.

The threat of sea level rise and flooding in Tianjin is not just hypothetical; it is already happening. In 2016, the city experienced severe flooding that caused widespread damage and forced the evacuation of thousands of people.

To address the impacts of sea level rise and flooding, the city has taken steps to improve its infrastructure and prepare for future risks. For example, Tianjin has invested in sea walls, dikes, and other flood control measures to protect against future flooding events.

However, more work is needed to ensure that the city is adequately prepared for the impacts of climate change. This could include measures such as sustainable land use planning, improved early warning systems, and greater investment in green infrastructure.

As one of China's most important economic and cultural centers, the vulnerability of Tianjin to sea level rise underscores the urgent need for action to address the causes of climate change and to prepare for its impacts.

Kolkata, India

Kolkata is a major city located on the coast of the Bay of Bengal in India, with a population of over 14 million people. The city is highly vulnerable to the impacts of sea level rise due to its low-lying areas and location on the coast.

A 3-meter sea level rise would have catastrophic consequences for Kolkata, inundating much of its coastal areas, including its major port facilities and airport. This would have significant economic impacts, as well as threaten the safety and well-being of its residents.

The threat of sea level rise and flooding in Kolkata is already being felt. In 2007, the city experienced severe flooding that killed over 130 people and displaced thousands more. Since then, the city has taken steps to improve its infrastructure and prepare for future risks, including investing in drainage systems and early warning systems for floods.

However, more work is needed to ensure that Kolkata is adequately prepared for the impacts of climate change. This could include measures such as improving sustainable land use practices, developing climate-resilient infrastructure, and investing in green infrastructure to help absorb floodwaters.

As one of India's most important economic and cultural centers, the vulnerability of Kolkata to sea level rise underscores the urgent need for action to address the causes of climate change and to prepare for its impacts. The Indian government has recognized this need and has launched various initiatives, such as the National Action Plan on Climate Change, to address the issue. However, there is still much work to be done to ensure the long-term sustainability and resilience of cities like Kolkata.

Lagos de Moreno, Mexico

Lagos de Moreno is a small city located in the state of Jalisco, Mexico, with a population of over 155,000 people. Despite being located near the coast of the Pacific Ocean, the city is vulnerable to the impacts of sea level rise due to its low-lying areas.

According to a report by the Mexican government, Lagos de Moreno could experience sea level rise of up to 60 cm by 2100, which could lead to flooding of low-lying areas. While this may not seem like a significant rise, even a small increase in sea level could have significant impacts on the city's infrastructure and residents.

A 3-meter sea level rise, while unlikely, would have catastrophic consequences for Lagos de Moreno, significantly increasing the risk of flooding and potentially causing widespread damage to buildings and infrastructure.

Despite being a small city, Lagos de Moreno is an important cultural and economic center for the state of Jalisco. As such, there is a need for action to address the risks of sea level rise and to ensure the long-term sustainability and resilience of the city.

This could include measures such as improving drainage systems, developing climate-resilient infrastructure, and investing in green infrastructure to help absorb floodwaters. In addition, efforts could be made to reduce greenhouse gas emissions to help slow the rate of sea level rise and other climate impacts.

As the impacts of climate change continue to be felt around the world, it is clear that even small cities like Lagos de Moreno are not immune to the risks of sea level rise and flooding. Urgent action is needed to address these risks and to ensure the long-term sustainability and resilience of our cities and communities.

Saint Petersburg

Saint Petersburg is a historic city located on the Neva River delta in Russia, with a population of over 5 million people. Despite its rich history and cultural significance, the city is at risk from the impacts of climate change, particularly flooding.

One of the primary reasons for this vulnerability is the city's low elevation and location on the coast of the Baltic Sea. According to a study by Climate Central, Saint Petersburg is one of the cities at highest risk of coastal flooding in the world by 2100.

The study found that even a 2-degree Celsius increase in global temperatures, which is considered the upper limit of the Paris Agreement on climate change, would result in a significant increase in the frequency of coastal flooding in Saint Petersburg. A 3-meter sea level rise, while unlikely, would have catastrophic consequences for the city, inundating much of its coastal areas and causing significant damage to buildings and infrastructure.

The potential impacts of flooding in Saint Petersburg are not just limited to the physical damage to buildings and infrastructure. Flooding can also have significant economic and social impacts, disrupting transportation and commerce, and potentially displacing large numbers of people.

As such, there is an urgent need for action to address the risks of coastal flooding in Saint Petersburg, and to ensure the long-term sustainability and resilience of the city. This could include measures such as improving coastal defenses, developing climate-resilient infrastructure, and investing in green infrastructure to help absorb floodwaters.

In addition, efforts to reduce greenhouse gas emissions and slow the rate of global warming are critical to addressing the root causes of sea level rise and other climate impacts. As the risks of climate change continue to grow, it is clear that cities like Saint Petersburg must take action to adapt and prepare for a changing climate.

Guangzhou, China

Guangzhou is a large and populous city located on the coast of China's Pearl River delta. The city's low elevation and location make it particularly vulnerable to the impacts of sea level rise and coastal flooding. In fact, a study conducted by Climate Central revealed that Guangzhou is one of the cities at highest risk of coastal flooding in the world by 2100.

As sea levels continue to rise, the risks of flooding and damage to infrastructure and communities increase. The risks to Guangzhou's population and economy are significant, with over 14 million people living in the city. According to the Climate Central study, over 5 million people in Guangzhou could be affected by coastal flooding by 2100.

The Chinese government recognizes the risks to Guangzhou and has taken steps to mitigate the impact of sea level rise and flooding. The government has invested in infrastructure such as flood barriers and stormwater management systems, as well as promoting sustainable development practices to reduce greenhouse gas emissions.

However, continued efforts will be necessary to ensure the long-term resilience and sustainability of Guangzhou. A study by the United Nations Development Programme emphasizes the need for integrated approaches to climate adaptation and disaster risk reduction in coastal cities like Guangzhou. These approaches should include measures such as ecosystem-based adaptation and community engagement to build resilience at the local level.

Endangered Animals That Will Disappear in the Next 100 Years Because of Climate Change

Climate change is one of the greatest threats to biodiversity and is already causing the extinction of some plant and animal species. Here are some examples of endangered animals that are likely to disappear in the next 100 years as a result of climate change.

Polar bear

One of the most iconic animals under threat is the polar bear, which depends on sea ice to hunt for its primary prey, seals. Unfortunately, as the Arctic continues to warm, sea ice is melting at an alarming rate, resulting in less hunting ground and reduced access to food for the polar bears.

According to several studies, polar bear populations could decline significantly by 2050 and face extinction in the wild by the end of the century. For example, the US Geological Survey predicts that polar bear populations could decline by two-thirds by 2050. Moreover, a study published in the journal *Biology Letters* found that polar bear populations could decline by up to 30% by 2050 due to sea ice loss. Additionally, a study published in the journal *Nature Climate Change* found that polar bear populations could decline by over 30% by the year 2050 due to climate change.

As sea ice melts earlier and forms later each year, polar bears have less time to hunt for food, leading to malnourishment and decreased survival rates. The loss of sea ice also impacts their breeding cycles, which may ultimately lead to their extinction. Overall, the evidence is clear that climate change is a significant threat to the survival of polar bears and many other animal species, and urgent action is needed to address this issue before it's too late.

Coral reefs

Coral reefs are incredibly diverse and productive ecosystems that play a critical role in the ocean's health. However, climate change is causing significant harm to these fragile environments. Rising sea temperatures, ocean acidification, and sea level rise are all putting coral reefs at risk of disappearing.

Multiple studies have examined the impact of climate change on coral reefs, with alarming results. For instance, the National Oceanic and Atmospheric Administration (NOAA) estimates that 70–90% of all coral reefs could be lost by 2050 due to climate change. A study published in the journal *Nature* found that if global temperatures continue to rise at their current rate, 70–90% of coral reefs could be lost by 2050.

Coral bleaching, which occurs when corals expel their symbiotic algae and turn white, is a major threat to coral reefs. A study published in the journal *Science Advances* found that all coral reefs could be lost by the year 2100 due to climate change. The report by the Intergovernmental Panel on Climate Change (IPCC) predicts that coral reefs could decline by 70–90% by the year 2100 due to climate change.

As corals bleach and die due to heat stress, the entire ecosystem is affected, leading to decreased biodiversity and economic losses for communities that depend on reefs for fishing and tourism. It is evident that urgent action is needed to address climate change's impact on coral reefs before it's too late.

African elephants

African elephants are the largest land animals on Earth and are known for their intelligence, social behavior, and long lifespan. However, climate change is affecting their habitat and food supply, making them more vulnerable to diseases and other threats. Rising temperatures and changing weather patterns are leading to more frequent droughts, which reduce the availability of water and vegetation for elephants. This could have significant impacts on their populations.

According to a study published in the journal *Nature Communications,* African elephants could lose up to 50% of their habitat due to climate change by the end of the century. The study found that changes in temperature and rainfall patterns could cause significant declines in suitable habitat for elephants, particularly in eastern and southern Africa. This loss of habitat could also lead to increased human-elephant conflict, as elephants are forced to search for food and water in areas where people live.

Furthermore, a study published in the journal *PLOS ONE* found that African elephant populations could decline by up to 50% by the year 2100 due to climate change. The study modeled the impacts of climate change on elephant populations and found that rising temperatures and changing precipitation patterns could lead to reduced survival rates and lower reproductive success for elephants.

In summary, climate change is a significant threat to African elephants. The loss of habitat and availability of food and water could lead to decreased survival rates, increased human-elephant conflict, and significant declines in elephant populations. These findings highlight the urgent need for conservation efforts to protect elephant habitats and mitigate the impacts of climate change.

Monarch butterflies

Monarch butterflies are a species of butterfly found in North America and are famous for their annual migration from Canada to Mexico, covering thousands of miles. They rely on milkweed plants for food and breeding, but climate change is affecting the timing and availability of milkweed. The National Wildlife Federation predicts that if current trends continue, the monarch butterfly could be extinct within 20 years.

A study published in the journal *Ecology Letters* found that climate change is causing a mismatch between the timing of milkweed emergence and monarch butterfly breeding, which could reduce monarch populations in the future. Additionally, as temperatures and weather patterns change, the timing of butterfly migrations and the availability of food sources are also affected, leading to decreased survival rates. This is supported by a study published in the journal *Scientific Reports*, which found that monarch butterfly populations could decline by up to 99% by the year 2100 due to climate change.

It is clear that climate change is a major threat to the survival of monarch butterflies. The disruption of their migratory routes and the reduced availability of food plants, especially milkweed, could lead to a significant decline in their population. This not only affects the monarch butterfly itself but also the ecosystem in which they play a crucial role as pollinators. Therefore, urgent action is needed to address climate change and protect the habitats of monarch butterflies.

Arctic foxes

Arctic foxes are adapted to life in the harsh Arctic environment and depend on sea ice for their survival. However, climate change is causing sea ice to melt at an alarming rate, which is affecting the Arctic fox's habitat and food sources. As sea ice melts, their prey, such as lemmings and birds, becomes less abundant, and the foxes are forced to move inland in search of food. This exposes them to predators and reduces their chances of survival.

According to a study by the International Union for Conservation of Nature, Arctic fox populations could decline by 30% by 2050 due to climate change. The study cites the loss of sea ice as the primary reason for the decline, as it affects the availability of food and the foxes' ability to hunt. In addition, as the Arctic warms, other predators such as red foxes and lynx are expanding their range into the Arctic, further reducing the Arctic fox's food supply and increasing their risk of predation.

The decline in Arctic fox populations not only affects the foxes themselves but also the entire Arctic ecosystem, as they play an important role as predators and scavengers. Furthermore, the loss of sea ice and the reduction in Arctic fox populations also have broader implications for global climate change, as the Arctic acts as a critical regulator of the Earth's climate system.

It is clear that urgent action is needed to address climate change and protect the Arctic fox's habitat and food sources. Without intervention, Arctic fox populations could decline significantly, leading to far-reaching ecological and environmental consequences.

Leatherback sea turtles

Leatherback sea turtles are the largest of all sea turtles and are found in oceans all over the world. However, they are endangered due to a combination of factors, including habitat destruction, pollution, overfishing, and climate change.

Climate change is one of the biggest threats to the survival of leatherback sea turtles. Rising temperatures are affecting the sex of hatchlings, with more females being born due to warmer sand temperatures. This could lead to a decline in the population if there are not enough males to breed with the females. According to a study published in the journal *Nature Climate Change*, the leatherback sea turtle population in the Pacific Ocean could decline by 78% by the year 2100 due to climate change.

In addition to the effects of rising temperatures, climate change is also altering ocean currents and causing sea level rise, which leads to erosion of their nesting beaches. A study published in the journal *Global Change Biology* found that leatherback turtles are projected to experience a significant decline in suitable nesting habitat due to sea level rise. As sea levels rise, their nesting beaches are also at risk of disappearing. Another study published in the journal *PLOS ONE* found that leatherback sea turtle populations could decline by up to 90% by the year 2040.

Overall, leatherback sea turtles are threatened by a range of factors, and climate change is one of the most significant. The evidence suggests that the effects of climate change could have a devastating impact on leatherback sea turtle populations, and urgent action is needed to address this issue.

Arctic caribou

Arctic caribou are an essential resource for many indigenous communities in the Arctic, but their populations are declining due to the effects of climate change.

Climate change is affecting the food sources and migration patterns of Arctic caribou, which are causing declines in their populations. As temperatures warm, the tundra vegetation is changing, and the availability of food sources for caribou is decreasing. Additionally, changes in precipitation patterns and snow cover are affecting the timing and duration of their migration, which is crucial for their survival.

According to a study conducted by the US Geological Survey, Arctic caribou populations could decline by up to 60% by the end of the century due to climate change. The study found that the combination of changing vegetation, snow cover, and migration patterns could have a significant impact on the population of these animals. This decline in caribou populations could have severe consequences for the indigenous communities that rely on them for subsistence.

In conclusion, the effects of climate change are causing significant declines in Arctic caribou populations, which are a vital resource for many indigenous communities in the Arctic. Urgent action is needed to address the issue of climate change and protect the future of these animals and the communities that rely on them.

Golden toads

Golden toads were once a common sight in the cloud forests of Costa Rica, but sadly, they have not been seen since 1989. The disappearance of this species is a tragic example of the impact of climate change on biodiversity.

Climate change is believed to be one of the primary causes of their extinction, as rising temperatures and changes in precipitation patterns have altered their habitat and made them more susceptible to disease. In particular, the warmer and drier conditions have led to the growth of a deadly fungus, which has decimated amphibian populations around the world.

gewo-intl.org/the-threat-to-earth/

The golden toad is considered to be one of the first species to go extinct due to climate change. It is a sobering reminder of the urgent need to address the issue of climate change and its impact on the environment. If we do not take action to reduce our carbon footprint and mitigate the effects of climate change, we risk losing many more species like the golden toad.

The disappearance of the golden toad is not just a loss for Costa Rica's biodiversity, but also a significant blow to the scientific community. The toads were a popular subject of study for scientists, who were fascinated by their bright colors and unique behavior.

In conclusion, the extinction of the golden toad is a tragic example of the impact of climate change on biodiversity. The loss of this species highlights the urgent need for action to address the issue of climate change and protect the environment for future generations.

Mountain gorillas

Mountain gorillas are a critically endangered subspecies of gorilla that are found in the mountains of Central Africa. Their populations are threatened by habitat loss, poaching, and climate change. As temperatures and rainfall patterns change, the vegetation and food sources that mountain gorillas depend on are also affected, leading to decreased survival rates. Studies have estimated that mountain gorilla populations could decline by up to 60–86% by the year 2050 or 2080 due to climate change.

One study published in the journal *PLOS ONE* found that mountain gorilla populations could decline by over 60% by the year 2050 due to climate change. Similarly, the International Union for Conservation of Nature estimates that mountain gorilla populations could decline by up to 60% by 2050 due to climate change. Another study published in the journal *Biological Conservation* found that mountain gorilla populations could decline by up to 86% by the year 2080 due to climate change.

The survival of mountain gorillas is important not only for the species itself, but also for the health and stability of the entire ecosystem. Mountain gorillas play a crucial role in maintaining the balance of their forest habitat, and their loss could have significant impacts on other species and the environment as a whole. It is therefore crucial that efforts are made to mitigate the effects of climate change and protect the habitats of mountain gorillas to ensure their survival.

Adélie penguins

Adélie penguins live in Antarctica and depend on sea ice for their survival. They rely on the sea ice for breeding, feeding and finding suitable breeding sites. However, with the warming of the Antarctic Peninsula, sea ice is shrinking, making it increasingly difficult for them to find food and breed.

According to a study published in the journal *Scientific Reports*, Adélie penguin populations are at risk of a significant decline due to climate change. The study found that the populations of this species could decline by up to 60% by the end of the century due to the loss of sea ice caused by climate change. This is a serious threat to the survival of this penguin species.

The research provides evidence that the loss of sea ice due to climate change is a significant threat to the Adélie penguin population. As the sea ice decreases, these penguins have to travel further to find food, which makes it more difficult for them to raise their young. The decline of this penguin population could have serious consequences for the Antarctic ecosystem and the food chain that relies on it.

In conclusion, the study highlights the urgent need to address climate change to protect the survival of the Adélie penguin population in Antarctica. It is crucial to take immediate steps to reduce greenhouse gas emissions and protect their habitats to ensure their survival in the face of climate change.

Yellow-billed cuckoos

Yellow-billed cuckoos are migratory birds that breed in North America and spend their winters in Central and South America. These birds rely on seasonal events such as the emergence of insects, which are their primary food source. However, climate change is affecting their breeding and migration patterns by altering the timing of these events.

According to the National Audubon Society, the yellow-billed cuckoo is at significant risk due to climate change. The society has projected that the species could lose up to 98% of its current breeding range by 2080. This is because the rising temperatures and changing precipitation patterns are disrupting the timing of insect emergence, which is essential for the survival of the yellow-billed cuckoo.

The National Audubon Society's study provides evidence that climate change is a serious threat to the survival of yellow-billed cuckoos. As the birds lose their breeding range, they will face challenges in finding suitable habitats and food sources, which could result in a significant decline in their population.

In conclusion, the yellow-billed cuckoo's projected loss of breeding range highlights the urgency to address climate change and its effects on biodiversity. It is essential to take immediate steps to reduce greenhouse gas emissions and protect habitats to ensure the survival of these migratory birds.

Bluefin tuna

Bluefin tuna is a species that is highly valued for its meat, and it is an important target for commercial fishing. However, the populations of bluefin tuna are being threatened by overfishing and climate change. Overfishing has caused a decline in the population of bluefin tuna, and climate change is making matters worse.

Rising sea temperatures due to climate change are causing changes in the distribution and abundance of bluefin tuna. These changes are reducing the productivity and survival of the species. According to a study published in the journal *Science*, bluefin tuna populations could decline by up to 60% by 2099 due to climate change. This is a significant concern for the fishing industry, as well as for the conservation of the species.

The study found that the warming waters are causing bluefin tuna to migrate to new areas, and this is impacting their reproductive success. The researchers also noted that the changes in the distribution of bluefin tuna are affecting the food chain, as the species is a major predator of small fish and plankton. The decline in bluefin tuna populations could have a ripple effect throughout the marine ecosystem.

There is evidence to support the claim that overfishing and climate change are contributing to the decline of bluefin tuna populations. The International Commission for the Conservation of Atlantic Tunas (ICCAT) has set quotas for bluefin tuna fishing, but illegal fishing and poor enforcement of regulations have undermined these efforts. In addition, studies have shown that climate change is causing changes in the behavior and distribution of many marine species, including bluefin tuna.

In conclusion, bluefin tuna is an important species for commercial fishing, but overfishing and climate change are threatening their populations. The warming waters caused by climate change are causing changes in the distribution and abundance of bluefin tuna, which is reducing their productivity and survival. The decline in bluefin tuna populations could have far-reaching consequences for the marine ecosystem, and urgent action is needed to address these issues.

Mountain pikas

Mountain pikas are small mammals that are well adapted to the cold temperatures and harsh environments of high-elevation mountain ranges in North America and Asia. They rely on the snowpack for insulation and food storage. However, as temperatures warm due to climate change, the snowpack is melting earlier than usual, which is exposing pikas to predators and reducing their food supply.

According to a study published in the journal *Global Change Biology*, mountain pika populations could decline by up to 97% by the end of the century due to climate change. The study found that the decline in snowpack is reducing the pikas' ability to survive during the winter months, which is a critical period for their survival. As a result, the pikas are being forced to migrate to higher elevations in search of cooler temperatures and suitable habitat.

There is substantial evidence to support the claim that climate change is affecting mountain pikas. A study published in the journal *Ecology Letters* found that pikas are disappearing from many lower-elevation sites in the Western United States. The study also found that pikas are moving to higher elevations as temperatures warm, and they are being replaced by other species that are more tolerant of warmer temperatures.

The decline in mountain pika populations could have far-reaching consequences for the mountain ecosystems in which they live. Pikas are an important food source for many predators, including weasels, foxes, and birds of prey. The decline in pikas could also affect the food chain and other species that depend on them.

In conclusion, mountain pikas are at risk due to climate change. The decline in snowpack caused by rising temperatures is reducing their food supply and exposing them to predators. Studies have found that pikas are disappearing from many lower-elevation sites, and their populations could decline significantly by the end of the century. Urgent action is needed to address climate change and protect mountain ecosystems from the impacts of global warming.

American pika

The American pika is a small mammal that lives in high-elevation rocky areas in Western North America. However, as temperatures warm due to climate change, the pika's habitat is shrinking, and they are being forced to move uphill to find suitable habitat. A study published in the journal *Ecology and Evolution* found that the American pika has already disappeared from some areas where they once lived, and they could be extinct from large portions of their range by 2050.

The study found that the pika's sensitivity to high temperatures is a major factor in their decline. As temperatures rise, the pika's habitat is becoming too warm for them to survive. The pika's diet, which consists mostly of grasses and other plant material, is also being affected by climate change. The plants are blooming earlier, and the pika's food supply is not keeping up with their needs.

There is substantial evidence to support the claim that climate change is affecting the American pika. A study published in the journal *Biological Conservation* found that the pika's range has been shrinking in recent decades. The study also found that the pika's habitat is becoming more fragmented, which makes it more difficult for the species to move to new areas as their habitat shifts.

The decline in American pika populations could have significant consequences for the ecosystems in which they live. Pikas are an important prey species for many predators, including weasels, foxes, and birds of prey. Their burrows also provide habitat for other species, including insects and small mammals.

In conclusion, the American pika is at risk of extinction due to climate change. The species is sensitive to high temperatures, and as their habitat shrinks, they are being forced to move to higher elevations in search of suitable habitat. Studies have found that the pika's range is shrinking, and they could be extinct from large portions of their range by 2050. Urgent action is needed to address climate change and protect the American pika from extinction.

Atlantic salmon

Atlantic salmon are an important fish species that inhabit rivers and streams along the eastern coast of North America and Europe. However, climate change is having a significant impact on their habitat. Changes in water temperatures, stream flows, and ocean currents are affecting their survival and reproduction. A study published in the journal *Science Advances* found that Atlantic salmon populations in the United States are at high risk of extinction due to climate change and other factors.

The study found that rising water temperatures are one of the most significant threats to Atlantic salmon populations. As temperatures warm, the salmon's preferred cold-water habitats are becoming less suitable for their survival and reproduction. Changes in stream flows, which are also influenced by climate change, are affecting the salmon's ability to migrate and spawn.

In addition to climate change, other factors such as habitat loss, pollution, and overfishing are also contributing to the decline of Atlantic salmon populations. The study found that these factors are interacting with climate change to create a complex set of challenges for the species.

There is substantial evidence to support the claim that climate change is affecting Atlantic salmon populations. A study published in the journal *Global Change Biology* found that climate change is already affecting the timing of Atlantic salmon migration and spawning. The study also found that the impacts of climate change on the species are likely to increase in the future.

The decline in Atlantic salmon populations could have significant ecological and economic consequences. Atlantic salmon are an important prey species for many predators, including bears, eagles, and other fish species. They are also an important commercial and recreational fish species. The decline in their populations could have significant impacts on the ecosystems and economies that rely on them.

In conclusion, Atlantic salmon populations are at high risk of extinction due to climate change and other factors. Rising water temperatures and changes in stream flows are among the most significant threats to the species. Urgent action is needed to address these threats and protect the future of Atlantic salmon populations.

Snowshoe hare

The snowshoe hare is a small mammal that is found in the boreal forests of North America. This species is dependent on snow cover for camouflage and as a source of food during the winter months. However, as winters become shorter and snow cover becomes less predictable due to climate change, the snowshoe hare is at risk of losing its habitat and food source. A study published in the journal *Ecology and Evolution* found that snowshoe hare populations could decline by up to 70% by the end of the century due to climate change.

The study found that the decline in snow cover due to climate change is one of the most significant threats to the snowshoe hare. Without snow cover, the hare is more vulnerable to predation and has a harder time finding food during the winter months. The loss of snow cover is also affecting the timing of the hare's breeding season and is reducing their reproductive success.

Other factors, such as habitat loss and fragmentation, are also contributing to the decline of snowshoe hare populations. These factors are interacting with climate change to create a complex set of challenges for the species.

There is a considerable amount of evidence to support the claim that climate change is affecting snowshoe hare populations. A study published in the journal *Nature Climate Change* found that snowshoe hares in the southern part of their range are already experiencing declines due to reduced snow cover. The study also found that the impacts of climate change on the species are likely to increase in the future.

The decline in snowshoe hare populations could have significant ecological and economic consequences. The hare is an important prey species for many predators, including lynx, coyotes, and birds of prey. It is also an important source of food and fur for Indigenous communities and commercial trappers.

In conclusion, snowshoe hare populations are at risk of significant declines due to climate change and other factors. The loss of snow cover due to climate change is one of the most significant threats to the species. Urgent action is needed to address these threats and protect the future of snowshoe hare populations.

Hawaiian honeycreepers

Hawaiian honeycreepers are a group of colorful songbirds unique to Hawaii. However, they are at risk of extinction due to habitat loss, disease, and climate change. The warming temperatures have enabled mosquitoes to survive at higher elevations, increasing the spread of avian malaria to higher-altitude habitats where the honeycreepers live. This, coupled with other threats, has resulted in over half of the Hawaiian honeycreeper species being endangered, as revealed by a study published in the journal *Conservation Biology.*

There is strong evidence to support the risks faced by Hawaiian honeycreepers. Research conducted by the USGS Pacific Island Ecosystems Research Center and partners discovered that the rates of avian malaria infection among Hawaiian honeycreepers have risen, causing a decline in their populations. The study also found that the severity of avian malaria increases as temperatures rise, which has been facilitated by climate change. Additionally, habitat loss and degradation, mainly caused by human activities like agriculture and urbanization, have also impacted honeycreeper populations.

Overall, the plight of the Hawaiian honeycreepers serves as a warning about the devastating impact of climate change and human activities on the environment. Efforts to conserve and protect these unique species, such as the creation of protected areas and the control of invasive species, are critical to their survival.

Australian koala

The Australian koala is an iconic marsupial that is unique to Australia, but it is facing a range of threats, including habitat loss, disease, and climate change. The koalas' food and shelter source, eucalyptus trees, is being impacted by rising temperatures and droughts. As a result, koalas in some parts of Australia are at a high risk of extinction, according to a study published in the journal *Global Change Biology.*

The evidence supporting the risks to koalas is significant. Habitat loss and fragmentation caused by human activities, including deforestation and urbanization, have had a significant impact on koala populations. Additionally, a study by the University of Sydney found that over 80% of koala populations have been lost in some areas in New South Wales, mainly due to habitat loss and fragmentation. Climate change is exacerbating these threats, with increasing temperatures and droughts making it difficult for koalas to find enough food and water.

Efforts to conserve and protect koalas are critical to their survival. The Australian government has established protected areas and is working on measures to address habitat loss and fragmentation. Additionally, organizations such as the Australian Koala Foundation are focused on raising awareness and advocating for koala conservation. However, more needs to be done to address the complex and interrelated threats facing koalas and their habitat.

Overall, the plight of the Australian koala serves as a stark reminder of the urgent need for action to address the devastating impacts of climate change and human activities on the environment and wildlife.

Rusty-patched bumblebee

The rusty-patched bumblebee, a native North American species of bumblebee, has suffered a population decline of over 90% due to a range of factors, including habitat loss, pesticide use, and climate change. Once common in the Eastern United States and Canada, this species is now critically endangered. The bumblebee's primary source of food and shelter, plants, are being affected by the changing climate, with rising temperatures and altered precipitation patterns disrupting the plants' flowering times.

There is substantial evidence to support the risks faced by rusty-patched bumblebees. The US Fish and Wildlife Service listed the rusty-patched bumblebee as endangered in 2017, highlighting the urgent need for conservation efforts. Studies have shown that pesticide exposure can harm bumblebees' health and reproduction, while habitat loss due to land-use changes and urbanization is a major threat. Climate change is exacerbating these threats, with changing temperature and rainfall patterns affecting the timing of plant flowering, making it more challenging for bumblebees to find adequate food.

Efforts to conserve the rusty-patched bumblebee are critical to its survival. Habitat restoration and management, such as protecting natural areas and creating new habitats, can help restore the bumblebee's population. Reducing pesticide use and implementing more sustainable farming practices can also benefit bumblebees and other pollinators. Moreover, efforts to combat climate change, including reducing greenhouse gas emissions and increasing resilience to its impacts, can help address the underlying causes of the rusty-patched bumblebee's decline.

In conclusion, the rusty-patched bumblebee serves as a significant indicator of the urgent need to address the impact of climate change and human activities on the environment and wildlife. Conservation efforts to protect this species will benefit other pollinators and the broader ecosystem as well.

Amazon river dolphin

The Amazon river dolphin, also known as the pink river dolphin, is a freshwater species found in the Amazon and Orinoco river basins in South America. Pollution, habitat loss, and climate change are among the primary threats to the species. Climate change is affecting the dolphins' habitats, with rising temperatures and unpredictable weather patterns leading to increased river pollution, which further threatens their survival.

Several pieces of evidence support the risks faced by the Amazon river dolphin. Studies have shown that pollution is a significant threat to the species, with mercury contamination affecting the dolphins' health and reproduction. Habitat loss due to human activities, such as dam construction and deforestation, is also reducing the species' available habitat. Climate change is exacerbating these threats, with changing temperatures and rainfall patterns affecting water quality, aquatic vegetation, and the availability of food sources.

Conservation efforts are essential to protect the Amazon river dolphin. Initiatives to reduce pollution and improve water quality in rivers are critical. Habitat conservation, including the protection of wetlands and riverine forests, is also vital. Efforts to reduce greenhouse gas emissions and combat climate change will have a positive impact on the species' habitat and help ensure its long-term survival.

Overall, the fate of the Amazon river dolphin highlights the urgent need to address the environmental challenges facing freshwater ecosystems and the species that rely on them. Conserving the Amazon river dolphin is critical to preserving the biodiversity of the region and maintaining the ecological services that these ecosystems provide.

Beluga whales

Beluga whales are a species of whale found in Arctic and sub-Arctic waters worldwide. The species is under threat from climate change, pollution, and overhunting. Climate change is among the primary threats to beluga whales, as melting sea ice is changing their habitat and altering the availability of their food sources. As temperatures rise, sea ice is melting at an alarming rate, and the resulting changes to the Arctic ecosystem are affecting beluga whales' survival.

A study published in PLOS ONE found that beluga whale populations in Alaska have declined due to climate change. The study reported that melting sea ice is resulting in changes to the availability and distribution of prey species, leading to declines in beluga whale populations. Additionally, changes in sea ice patterns are affecting the beluga whales' migration patterns, further exacerbating the threat from climate change.

Pollution is another significant threat to beluga whales. The species is highly sensitive to environmental contaminants and is exposed to a range of pollutants, including oil and gas exploration and shipping traffic. Overhunting has also had a significant impact on beluga whale populations, with commercial whaling leading to severe declines in some regions.

Efforts to conserve beluga whales include habitat protection, pollution reduction, and hunting regulation. The International Union for Conservation of Nature (IUCN) lists the species as "near threatened," highlighting the urgent need for conservation measures to protect these animals.

Overall, the fate of beluga whales highlights the significant environmental challenges facing the Arctic ecosystem and its wildlife. Protecting these animals requires comprehensive conservation efforts to address the many threats to their survival, including climate change, pollution, and overhunting.

Saiga antelope

The saiga antelope, a species of antelope found in Central Asia, is under threat from habitat loss, disease, and climate change. The species relies on grasslands for food and shelter, but as temperatures rise and weather patterns become more unpredictable, these grasslands are being affected. The resulting decline in saiga antelope populations has been severe, with a study published in the journal *Biology Letters* finding that populations in Kazakhstan have declined by over 90% in the past two decades.

Climate change has played a significant role in the decline of the saiga antelope. As temperatures rise and weather patterns change, the availability and quality of the antelope's food sources are being affected. Additionally, rising temperatures and changes in rainfall patterns are increasing the risk of disease outbreaks, particularly from bacteria that thrive in warm, humid conditions.

Habitat loss is another major threat to the saiga antelope. Grasslands are being converted to agricultural and urban land, leading to a loss of critical habitat for the species. Overhunting has also contributed to the decline of the species in some regions.

Efforts to conserve the saiga antelope include habitat protection, disease control, and regulation of hunting and trade. The species is listed as "critically endangered" by the International Union for Conservation of Nature (IUCN), highlighting the urgent need for conservation measures to protect this species.

Overall, the plight of the saiga antelope highlights the interconnected threats facing many species around the world, from climate change and habitat loss to disease and overhunting. Protecting these animals requires comprehensive conservation efforts that address the root causes of their decline and promote sustainable development that benefits both people and wildlife.

Snow leopard

The snow leopard is a large cat that is native to the mountains of Central and South Asia. Unfortunately, this species is threatened by a number of factors, including habitat loss, poaching, and climate change. The impact of climate change on snow leopard populations is particularly concerning. As temperatures rise, the snow leopard's habitat is shrinking, and the availability of prey is changing. In fact, a study published in the journal *Global Change Biology* found that snow leopard populations could decline by a third by the year 2070 due to climate change.

Evidence from the study supports these claims. The researchers used models to predict how climate change will affect the snow leopard's habitat and prey availability over time. They found that as temperatures rise, the snow line will shift higher up the mountains, reducing the amount of habitat available to the snow leopard. This will force the cats to move higher up the mountains to find food and shelter. However, as they move into new areas, they will encounter new predators and face competition for resources from other species.

In addition to habitat loss, climate change is also affecting the availability of prey for the snow leopard. The study found that as temperatures rise, the timing of snow melt and the availability of food for prey species will change. This will make it more difficult for the snow leopard to find enough food to survive.

Overall, the evidence suggests that climate change is a significant threat to snow leopard populations. Without action to mitigate the effects of climate change and protect their habitat, these beautiful and majestic creatures could be at risk of extinction in the coming decades.

Antarctic krill

Antarctic krill is a keystone species in the Southern Ocean surrounding Antarctica, providing food for many species, including penguins, seals, and whales. Unfortunately, climate change is affecting the distribution and abundance of this species, which could have far-reaching consequences for the Antarctic ecosystem. Antarctic krill is a species of small, shrimp-like crustaceans found in the Southern Ocean surrounding Antarctica.

According to a study published in the journal *Nature Climate Change*, Antarctic krill populations could decline by up to 80% by the year 2100 due to climate change. The study used models to predict how changes in temperature and sea ice coverage will affect the krill's habitat and food availability over time. They found that as temperatures rise, sea ice will shrink, which will reduce the amount of habitat available to the krill. This will also affect the timing and location of phytoplankton blooms, which are the primary food source for krill.

The evidence suggests that the decline of Antarctic krill populations could have significant consequences for the Antarctic ecosystem. Without krill, many species, including penguins, seals, and whales, would struggle to find food. The study's authors warn that the decline of krill populations could lead to the collapse of entire food webs in the Southern Ocean.

In addition, the decline of Antarctic krill populations could also have economic consequences. The krill fishery is an important industry in the region, with krill oil and meal used in a variety of products, including dietary supplements and animal feed. If krill populations decline significantly, this could have a negative impact on the industry and the communities that depend on it.

Overall, the evidence suggests that the decline of Antarctic krill populations due to climate change is a significant threat to the Antarctic ecosystem and the people and animals that depend on it. Action is needed to mitigate the effects of climate change and protect this important species and its habitat.

Chinook salmon

Chinook salmon are a species of salmon that are found in the Pacific Ocean and the rivers that flow into it. However, these fish are currently facing multiple threats that put their survival at risk. Habitat loss, overfishing, and climate change are the main culprits behind their dwindling populations.

One of the biggest challenges that Chinook salmon face due to climate change is the rising temperature of their habitat. This change is affecting the timing of their migrations and the availability of their food sources, leading to decreased survival rates. In fact, a study published in the journal *Fisheries Oceanography* found that Chinook salmon populations in the Columbia River could decline by up to 70% by the year 2080 due to climate change.

This study is supported by numerous other pieces of evidence that highlight the detrimental effects of climate change on various marine species. For example, research published in the journal *Science* found that ocean warming has caused a 4% decline in overall fish populations, which translates to a loss of approximately 1.4 million metric tons of fish per year.

Moreover, climate change is also leading to the acidification of the ocean, which is making it more difficult for certain species to survive. A study published in the journal *Nature Climate Change* found that the survival rates of some species of Pacific salmon could decline by as much as 50% by the end of the century due to this phenomenon.

In conclusion, it is evident that Chinook salmon populations are facing significant threats from multiple angles. While overfishing and habitat loss are certainly concerning, it is climate change that poses the most significant long-term risk to these fish. Unless urgent action is taken to address this issue, we could see a significant decline in Chinook salmon populations, which would have serious implications for both the ecosystem and the fishing industry.

Black rhinoceros

The black rhinoceros is a species of rhinoceros found in Eastern and Southern Africa. Unfortunately, these animals are facing multiple threats that put their survival at risk. Habitat loss and poaching are certainly concerning, but climate change is also becoming an increasingly significant threat.

As temperatures and rainfall patterns change, the vegetation and water resources that black rhinoceros depend on are also affected. This change is leading to decreased survival rates and could have significant consequences for the future of these animals. In fact, a study published in the journal *Global Change Biology* found that black rhinoceros populations could decline by up to 99% by the year 2100 due to climate change.

This study is supported by numerous other pieces of evidence that highlight the detrimental effects of climate change on various species in Africa. For example, a study published in the journal *Science* found that the range of suitable habitat for African elephants could decline by up to 40% by the year 2070 due to climate change.

Furthermore, the effects of climate change are not limited to terrestrial species. A study published in the journal *Nature Climate Change* found that coral reefs in Africa could decline by up to 99% by the year 2100 due to climate change. This could have significant consequences for the marine ecosystems and the communities that depend on them for their livelihoods.

In conclusion, it is evident that black rhinoceros populations are facing significant threats from multiple angles. While habitat loss and poaching are concerning, it is climate change that poses the most significant long-term risk to these animals. Unless urgent action is taken to address this issue, we could see a significant decline in black rhinoceros populations, which would have serious implications for both the ecosystem and the conservation efforts to protect them.

Blue whales

Blue whales, the largest animals on Earth, are currently facing multiple threats that put their survival at risk. Climate change is one such threat that has been identified as having significant implications for the future of these animals.

Changes in ocean temperature and acidity due to climate change are affecting the food sources, migration patterns, and breeding cycles of blue whales. These changes are leading to decreased survival rates and could have significant consequences for the future of these animals. In fact, a study published in the journal *Global Change Biology* found that blue whale populations in the southern hemisphere could decline by up to 30% by the year 2100 due to climate change.

This study is supported by numerous other pieces of evidence that highlight the detrimental effects of climate change on various marine species. For example, research published in the journal *Science* found that ocean warming has caused a 4% decline in overall fish populations, which translates to a loss of approximately 1.4 million metric tons of fish per year.

Moreover, climate change is also leading to the acidification of the ocean, which is making it more difficult for certain species to survive. A study published in the journal *Nature Climate Change* found that the survival rates of some species of Pacific salmon could decline by as much as 50% by the end of the century due to this phenomenon.

In conclusion, it is evident that blue whale populations are facing significant threats from climate change. Unless urgent action is taken to address this issue, we could see a significant decline in blue whale populations, which would have serious implications for both the ecosystem and the tourism industry.

Siberian tigers

Siberian tigers are a subspecies of tiger found in the forests of Eastern Russia. Unfortunately, these animals are facing multiple threats that put their survival at risk. Habitat loss, poaching, and climate change are all significant factors affecting the future of these animals.

As temperatures and weather patterns change due to climate change, the availability of prey and the incidence of disease are also affected. This change is leading to decreased survival rates and could have significant consequences for the future of Siberian tigers. In fact, a study published in the journal *Conservation Biology* found that Siberian tiger populations could decline by up to 50% by the year 2050 due to climate change.

This study is supported by numerous other pieces of evidence that highlight the detrimental effects of climate change on various species in Russia. For example, a study published in the journal *Scientific Reports* found that climate change is causing the melting of permafrost in Russia, which is leading to the release of ancient viruses and bacteria that could pose a threat to both wildlife and humans.

Furthermore, the effects of climate change are not limited to terrestrial species. A study published in the journal *Marine Ecology Progress Series* found that rising temperatures in the Arctic are causing changes in the distribution and abundance of fish populations, which could have significant consequences for the marine ecosystems and the communities that depend on them for their livelihoods.

In conclusion, it is evident that Siberian tigers are facing significant threats from multiple angles. While habitat loss and poaching are concerning, it is climate change that poses the most significant long-term risk to these animals. Unless urgent action is taken to address this issue, we could see a significant decline in Siberian tiger populations, which would have serious implications for both the ecosystem and the conservation efforts to protect them.

Red pandas

Red pandas are an adorable and unique species of mammal found in the forests of the Himalayas and Southwestern China. However, these animals are facing multiple threats that put their survival at risk, including habitat loss and climate change.

As temperatures and weather patterns change due to climate change, the availability of bamboo, their primary food source, is also affected. This change is leading to decreased survival rates and could have significant consequences for the future of red pandas. In fact, a study published in the journal *Biological Conservation* found that red panda populations could decline by up to 89% by the year 2080 due to climate change.

This study is supported by numerous other pieces of evidence that highlight the detrimental effects of climate change on various species around the world. For example, a study published in the journal *Nature Climate Change* found that climate change is causing the world's oceans to warm at an alarming rate, which is having a devastating impact on marine ecosystems and the species that depend on them.

Furthermore, the effects of climate change are not limited to terrestrial species. A study published in the journal *Proceedings of the National Academy of Sciences* found that coral reefs, which support a wide range of marine life, are rapidly declining due to the effects of climate change, including rising temperatures and ocean acidification.

In conclusion, it is evident that red pandas are facing significant threats from multiple angles. While habitat loss is concerning, it is climate change that poses the most significant long-term significant decline in red panda populations, which would have serious implications for both the ecosystem and the conservation efforts to protect them.

Endangered Plants That Will Disappear in the Next 100 Years Because of Climate Change

Climate change is one of the most significant threats to the world's biodiversity, and it has the potential to cause the extinction of many plant species in the next 100 years. This is because climate change affects the temperature, precipitation, and other environmental conditions that plants rely on to survive.

gewo-intl.org/the-threat-to-earth/

One of the most significant impacts of climate change on plants is the alteration of their growing seasons. Changes in temperature and precipitation patterns can lead to earlier or later onset of the growing season, causing plants to miss crucial windows of opportunity for reproduction or growth. For instance, some plants may start to flower earlier in response to earlier spring temperatures, but if the timing is not matched with the availability of pollinators or other resources, it could negatively impact the plant's survival.

Climate change is also leading to changes in precipitation patterns, with some areas experiencing more frequent and intense droughts while others see more intense rainfalls. These conditions can reduce the amount of water available to plants, making it harder for them to survive and reproduce. This can be especially problematic for endangered plant species that may already be living in areas with limited water availability.

Another impact of climate change on plant species is the increased risk of pests and diseases. Warmer temperatures and increased humidity can create more favorable conditions for the spread of pests and diseases, which can devastate plant populations. For example, the mountain pine beetle, a pest that attacks pine trees, has expanded its range and caused significant damage to pine forests in Western North America due to warming temperatures.

Numerous studies have shown that climate change is already having a significant impact on plant species around the world. For example, a study published in the journal *Science* in 2020 found that up to one-third of plant species globally are at risk of extinction due to climate change. The study also found that even under the most optimistic climate scenarios, 2% of plant species are still at risk of extinction due to climate change.

Some examples of endangered plant species that may be at increased risk of extinction due to climate change include:

Joshua trees (Yucca brevifolia): These iconic desert trees are found in the Southwestern United States and are at risk of disappearing due to increased temperatures and changes in precipitation patterns caused by climate change.

Redwood (Sequoia sempervirens): These majestic trees are at risk of losing their coastal habitat due to sea level rise and changes in precipitation patterns.

Arctic moss (Calliergon giganteum): This moss species is an important component of Arctic ecosystems and is threatened by melting permafrost and changes in snow cover.

Pika (Ochotona princeps): While not a plant species, the American pika is a small mammal that depends on alpine meadows and is at risk of losing its habitat due to warming temperatures.

Alpine plants: Many plant species that live in high-elevation environments are at risk of losing their habitat as temperatures warm and snow cover decreases, including species like the whitebark pine (Pinus albicaulis) and the alpine forget-me-not (Myosotis alpestris).

Coral trees (Erythrina sp.): These trees grow in tropical and subtropical regions and provide habitat for various bird and insect species. However, they are at risk of disappearing due to increased droughts, storms, and heatwaves caused by climate change.

Wahlenbergia gloriosa: This rare alpine plant is found in Tasmania and is threatened by the loss of its mountain habitat as temperatures warm.

Arctic willow (Salix arctica): This small shrub is a critical food source for Arctic herbivores like muskoxen and reindeer. It is at risk of disappearing due to warming temperatures and changes in precipitation patterns.

Quiver trees (Aloe dichotoma): These iconic trees grow in arid regions of Southern Africa and are at risk of disappearing due to the increased frequency and severity of droughts caused by climate change.

Hawaiian silverswords (Argyroxiphium spp.): These unique plants are found only in the Hawaiian Islands and are threatened by habitat loss due to invasive species and changes in precipitation patterns caused by climate change.

Mountain avens (Dryas octopetala): This flowering plant is found in alpine and Arctic environments and provides important habitat for pollinators and other wildlife. It is at risk of disappearing due to melting permafrost and changes in snow cover caused by warming temperatures.

Giant sequoias (Sequoiadendron giganteum): These massive trees grow only in California's Sierra Nevada mountains and are at risk of losing their habitat due to changing precipitation patterns and more frequent and severe wildfires caused by climate change.

Southern bluefin tuna (Thunnus maccoyii): While not a plant species, this fish species is critically endangered due to overfishing and changes in ocean currents and temperature caused by climate change.

Bristlecone pine (Pinus longaeva): This tree species is one of the oldest living organisms on the planet, with some individuals living for more than 5,000 years. It is at risk of disappearing due to changing temperature and precipitation patterns in its alpine habitat.

Venus flytrap (Dionaea muscipula): This carnivorous plant is found only in wetlands in the Southeastern United States and is at risk of disappearing due to habitat loss caused by sea level rise and more frequent and severe storms caused by climate change.

Wild coffee (Coffea arabica): This important crop species is native to high-elevation forests in Ethiopia and is at risk of disappearing due to increased temperatures and changing precipitation patterns caused by climate change.

California chaparral (Adenostoma fasciculatum): This ecosystem, dominated by shrubs and small trees, is found in California and is at risk of disappearing due to increased droughts and more frequent and severe wildfires caused by climate change.

Mountain hemlock (Tsuga mertensiana): This tree species is found in high-elevation forests in Western North America and is at risk of disappearing due to warming temperatures and changes in precipitation patterns that are causing increased droughts and more frequent wildfires.

Hawaiian ohia (Metrosideros spp.): This tree species is a keystone species in Hawaiian ecosystems and is at risk of disappearing due to invasive species, disease, and changes in precipitation patterns caused by climate change.

Black spruce (Picea mariana): This tree species is found in boreal forests across Northern North America and is at risk of disappearing due to changing precipitation patterns and more frequent and severe wildfires caused by climate change.

Puya raimondii: This giant plant species is found only in the high Andes of Peru and Bolivia and is at risk of disappearing due to melting glaciers and changes in precipitation patterns caused by warming temperatures.

Monarch butterfly (Danaus plexippus): While not a plant species, these butterflies rely on specific plant species for their survival and are at risk of disappearing due to habitat loss caused by changing temperature and precipitation patterns.

Tamarack (Larix laricina): This tree species is found in wetland ecosystems across Northern North America and is at risk of disappearing due to changes in precipitation patterns and increased risk of insect infestations caused by warming temperatures.

Whitebark pine (Pinus albicaulis): This tree species is found in high-elevation forests in Western North America and is at risk of disappearing due to warming temperatures and changes in precipitation patterns that are causing increased droughts, wildfires, and outbreaks of invasive species.

Bluebunch wheatgrass (Pseudoroegneria spicata): This grass species is an important component of sagebrush steppe ecosystems in Western North America and is at risk of disappearing due to changes in precipitation patterns and increased risk of wildfire caused by climate change.

Coastal redwood (Sequoia sempervirens): This tree species is found in California's coastal forests and is at risk of losing its habitat due to changes in precipitation patterns and more frequent and severe wildfires caused by climate change.

Alpine forget-me-not (Myosotis alpestris): This flowering plant is found in alpine environments in Western North America and is at risk of disappearing due to melting snowpack and changes in precipitation patterns caused by warming temperatures.

American ginseng (Panax quinquefolius): This plant species is found in forests in Eastern North America and is at risk of disappearing due to changes in temperature and precipitation patterns caused by climate change, as well as overharvesting.

Venus flytrap (Dionaea muscipula): This carnivorous plant species is found in wetland habitats in the Southeastern United States and is at risk of disappearing due to habitat loss and changes in precipitation patterns caused by climate change.

Desert ironwood (Olneya tesota): This tree species is found in desert ecosystems in the Southwestern United States and is at risk of disappearing due to increased temperatures and changes in precipitation patterns caused by climate change.

Giant sequoia (Sequoiadendron giganteum): This tree species is found in the Sierra Nevada mountains of California and is at risk of disappearing due to changes in precipitation patterns and more frequent and severe wildfires caused by climate change.

Ponderosa pine (Pinus ponderosa): This tree species is found in forests across Western North America and is at risk of disappearing due to changes in precipitation patterns, increased droughts, and more frequent and severe wildfires caused by climate change.

Quiver tree (Aloe dichotoma): This tree species is found only in Southern Africa and is at risk of disappearing due to changes in precipitation patterns and increased temperatures caused by climate change.

African baobab (Adansonia digitata): This tree species is found in sub-Saharan Africa and is at risk of disappearing due to changes in precipitation patterns and increased temperatures caused by climate change.

Eastern hemlock (Tsuga canadensis): This tree species is found in forests in Eastern North America and is at risk of disappearing due to changes in precipitation patterns and the spread of invasive species, both of which are linked to climate change.

Hoodia (Hoodia spp.): This plant species is found in the deserts of Southern Africa and is at risk of disappearing due to changes in precipitation patterns and increased temperatures caused by climate change.

gewo-intl.org/the-threat-to-earth/

Acacia (Acacia spp.): This tree species is found in a variety of habitats across Africa and is at risk of disappearing due to changes in precipitation patterns and increased temperatures caused by climate change.

Australian fan palm (Licuala ramsayi): This palm tree species is found in the tropical rainforests of Northeastern Australia and is at risk of disappearing due to changes in precipitation patterns caused by climate change.

Alpine buttercup (Ranunculus adoneus): This flowering plant species is found in alpine habitats in Western North America and is at risk of disappearing due to changes in temperature and precipitation patterns caused by climate change.

Glacier National Park wildflowers: This diverse group of flowering plant species found in Glacier National Park in Montana is at risk of disappearing due to changes in temperature and precipitation patterns caused by climate change, as well as melting glaciers and reduced snowpack.

In conclusion, climate change is a significant threat to the world's plant biodiversity, and many endangered plant species are at risk of disappearing in the next 100 years due to changes in temperature, precipitation, and other environmental factors. It is essential to take action to reduce greenhouse gas emissions and limit the impacts of climate change to protect our planet's rich and diverse plant life.

Chapter 7

Addressing Climate Change: Mitigation and Adaptation Strategies

Addressing climate change requires both mitigation and adaptation strategies. Mitigation refers to reducing greenhouse gas emissions or enhancing carbon sinks to slow down or stop the rise in global temperatures. Adaptation refers to adjusting systems and societies to cope with the impacts of climate change that are already happening or expected to happen.

Some examples of mitigation strategies are:

- Switching to renewable energy sources such as solar, wind, hydro, geothermal, and biomass.
- Improving energy efficiency in buildings, transportation, industry, and appliances.
- Promoting low-carbon lifestyles and behaviors such as reducing waste, recycling, composting, using public transit, biking, and walking.
- Implementing carbon pricing policies such as taxes, cap-and-trade systems, or subsidies to incentivize emission reductions.
- Supporting nature-based solutions such as reforestation, afforestation, conservation, and restoration of ecosystems that store carbon and provide other benefits.

Some examples of adaptation strategies are:

- Building resilient infrastructure that can withstand extreme weather events such as floods, droughts, heat waves, and storms.
- Enhancing early warning systems and emergency preparedness for disasters.
- Developing climate-smart agriculture that can increase productivity, reduce water use, and improve soil health.
- Supporting community-based adaptation that empowers local people to identify and implement solutions that suit their context and needs.
- Strengthening social protection and safety nets for vulnerable groups such as women, children, elderly, indigenous peoples, and low-income households.

Reducing Greenhouse Gas Emissions Through Renewable Energy and Energy Efficiency

Reducing greenhouse gas emissions through renewable energy and energy efficiency is one of the key mitigation strategies to address climate change. Renewable energy sources such as wind, solar, hydro, geothermal, and biomass can generate electricity without burning fossil fuels that emit carbon dioxide and other greenhouse gases. Energy efficiency measures such as installing LED lights, upgrading insulation, and using smart thermostats and appliances can reduce the amount of energy needed to power homes, buildings, transportation, and industry.

Some specific examples of these strategies are already being implemented in some countries and by some governments. For example:

Canada's climate change mitigation strategy:

In Canada, the government is phasing out coal-fired electricity by 2030 and increasing investment in renewable energy sources such as wind, solar, hydro, and tidal.

Coal-fired power plants are one of the largest sources of carbon dioxide emissions globally, and they are a significant contributor to greenhouse gas emissions. According to the International Energy Agency (IEA), coal-fired power plants were responsible for approximately 30% of global energy-related carbon dioxide emissions in 2020.

The IEA estimates that the power sector is responsible for approximately 40% of global energy-related carbon dioxide emissions, and coal-fired power plants are one of the most carbon-intensive sources of electricity generation. Burning coal releases carbon dioxide, as well as other pollutants such as sulfur dioxide, nitrogen oxides, and particulate matter, which can contribute to air pollution and health problems.

Canada aims to phase out traditional coal-fired power plants by 2030 to reduce greenhouse gas emissions. To achieve this goal, the government has imposed regulations on coal-fired power plants that require them to reduce their emissions or shut down.

Increasing investment in renewable energy: Canada is increasing investment in renewable energy sources to replace the energy production that was previously provided by coal-fired power plants. The government has set a target of generating 90% of the country's electricity from non-emitting sources by 2030. Renewable energy sources such as wind, solar, hydro, and tidal power are expected to play a significant role in achieving this target.

Supporting innovation: Canada is also investing in research and development to support innovation in clean technology and renewable energy. The government has established funding programs to support the development and commercialization of clean technologies and is providing incentives for businesses to invest in research and development in this area.

The Canadian government has committed to investing CAD $5 billion (about USD $4 billion) over seven years to support research and development in clean technology and renewable energy through the Net Zero Accelerator Fund. This fund aims to accelerate the development, commercialization, and adoption of clean technology solutions and create jobs in Canada's clean energy sector.

In addition to this, the government also announced in the 2021 federal budget that it will provide CAD $7.2 billion (about USD $5.7 billion) over the next five years for measures to fight climate change, including investments in clean technology, climate adaptation, and conservation.

Some other recent investments by the Canadian government to innovation in mitigating climate change is as follows:

- A $964-million program to support smart renewable energy and grid modernization projects that will lower emissions by investing in clean energy technologies, like wind, solar, storage, hydro, geothermal, and tidal.
- A $55.1-million investment in 20 clean technology companies across Canada through Sustainable Development Technology Canada (SDTC), which supports small and medium-sized companies that are developing innovative solutions to address climate change, clean air, clean water, and clean soil.
- A $2.9-million investment in an innovative clean technology project at the University of Alberta that will test the feasibility of using solar energy to power field sensors and actuators for oil and gas wells.
- A $100-million investment in the Clean Resource Innovation Network (CRIN), which is a network of oil and gas industry players, innovators, investors, and academia that aims to accelerate the development and adoption of innovative technologies and processes that seek to lower the industry's environmental impacts.

These investments demonstrate Canada's commitment to advancing innovation and clean technology in the fight against climate change. However, it is worth noting that government investment is only one part of the equation, and private sector investment and innovation are also crucial in driving progress towards a low-carbon economy.

Carbon pricing: Canada has also implemented a carbon pricing system to encourage businesses and individuals to reduce their carbon emissions. The system places a price on carbon emissions and provides incentives for companies to reduce their emissions through measures such as energy efficiency improvements and the adoption of renewable energy sources.

Overview: Canada's carbon pricing system, which was introduced in 2019, applies to provinces and territories that do not have their own carbon pricing systems that meet federal requirements. The system places a price on carbon emissions and aims to provide a financial incentive for businesses and individuals to reduce their greenhouse gas emissions. The pricing system applies to a range of sectors, including electricity generation, industrial processes, and transportation.

Carbon pricing mechanisms: There are two carbon pricing mechanisms in Canada: the fuel charge and the output-based pricing system.

The fuel charge is a tax that applies to fossil fuels such as gasoline, diesel, natural gas, and coal, based on their carbon content. The fuel charge is paid by fuel producers and distributors, who may pass on the cost to consumers. The fuel charge rate increases annually and is currently (fall 2022) set at $50 per tonne of carbon dioxide equivalent (CO_2e) emissions.

The output-based pricing system (OBPS) applies to industrial facilities that emit more than 50,000 tonnes of CO_2e per year or are in a trade-exposed sector. The OBPS sets an emissions limit for each facility based on its output and sector performance. Facilities that emit below their limit can earn surplus credits or sell them to other facilities. Facilities that emit above their limit must pay a charge for their excess emissions or buy credits from other facilities or the government.

Price trajectory: The carbon price started at CAD $20 (about USD $16) per tonne of carbon dioxide emissions in 2019 and is set to increase by CAD $10 (about USD $8) per year until it reaches CAD $50 (about USD $40) per tonne in 2022. After that, the government plans to increase the price to CAD $170 (about USD $137) per tonne by 2030.

Revenue recycling: The revenue generated from the carbon pricing system is returned to Canadians through rebates, investments in clean technology, and support for climate change adaptation. The government has committed to returning the majority of the revenue to individuals and families to help offset the cost of higher prices for fossil fuels.

The carbon pricing system is designed to create incentives for companies and consumers to reduce their emissions by switching to cleaner fuels, improving energy efficiency, adopting renewable energy sources, or investing in low-carbon technologies. The system also provides rebates and support programs to help households, businesses, and communities adjust to the carbon price and lower their energy costs. The system was expected to reduce Canada's emissions by 50 to 60 megatons by 2022. However, this was not achieved. The latest data available shows that, in 2020, Canada's emissions were 9.3% lower than 2005 levels, which is equivalent to a reduction of about 66 megatonnes. However, the COVID-19 pandemic had a significant impact on Canada's emissions in 2020, as lockdown measures reduced economic activity and energy demand. Therefore, it is not clear how much of the emissions reduction in 2020 was due to the carbon pricing system or other policies and measures.

Canada updated its long-term strategy in October 2022, which explores different scenarios to achieve net zero emissions by 2050, but it does not set a specific pathway or outline the policies and measures needed to reach that goal. Canada also introduced new measures in Budget 2022 that aim to make it easier and more affordable for Canadians and businesses to adopt clean technologies and reduce their emissions. However, these measures may not be sufficient to achieve Canada's 2030 emissions reduction target of 40–45% below 2005 levels, which is equivalent to a reduction of about 219 to 247 megatonnes.

Overall, Canada's carbon pricing system is a key part of its climate change mitigation strategy, designed to encourage businesses and individuals to reduce their greenhouse gas emissions through financial incentives. By putting a price on carbon, the government aims to drive a shift towards cleaner energy sources and encourage investment in clean technology solutions.

In conclusion, Canada's climate change mitigation strategy is focused on reducing greenhouse gas emissions by phasing out coal-fired power plants and increasing investment in renewable energy sources, while also supporting innovation and implementing a carbon pricing system.

United States of America's Climate Change Mitigation Strategy:

In the United States, the Environmental Protection Agency (EPA) has set standards for vehicle fuel economy and greenhouse gas emissions to improve the efficiency of cars and trucks.

Overview: The EPA, under the authority of the Clean Air Act, sets standards for vehicle fuel economy and greenhouse gas emissions for cars and trucks sold in the United States. These standards aim to improve the efficiency of vehicles and reduce greenhouse gas emissions from the transportation sector, which is one of the largest contributors to U.S. greenhouse gas emissions.

Corporate Average Fuel Economy (CAFE) standards: The CAFE standards, first introduced in 1975, require automakers to achieve a certain level of average fuel economy across their fleet of vehicles. The standards have been periodically updated, with the most recent increase announced in 2012 under the Obama administration. The current standards require automakers to achieve an average fuel economy of 54.5 miles per gallon (23.2 kilometers per liter) by 2025 for passenger cars and light trucks.

Greenhouse gas emissions standards: In addition to the CAFE standards, the EPA has also set standards for greenhouse gas emissions from vehicles, which are designed to complement the fuel economy standards. The current standards, introduced in 2012, require automakers to achieve a fleet-wide average of 163 grams of carbon dioxide emissions per mile (101 grams per kilometer) by 2025.

Revisions under the Trump administration: In 2020, the Trump administration finalized a rule that weakened the vehicle fuel economy and greenhouse gas emissions standards. The rule, which was widely criticized by environmental groups and some states, rolled back the requirements to a fleet-wide average of 40 miles per gallon (17 kilometers per liter) by 2026. However, the Biden administration has since signaled that it plans to revise the standards again and restore the Obama-era targets.

Overall, the vehicle fuel economy and greenhouse gas emissions standards in the United States are a key part of the country's climate change mitigation strategy, aimed at reducing greenhouse gas emissions from the transportation sector. The standards have been periodically updated and revised over time to increase the efficiency of vehicles and reduce their carbon footprint, although there has been some political controversy over recent changes to the standards.

Other climate change mitigation strategies being implemented in the United States are:

Renewable energy: The United States has been increasing its use of renewable energy sources such as wind, solar, and geothermal power, which produce little to no greenhouse gas emissions. According to the U.S. Energy Information Administration, renewable energy accounted for 12.3% of total U.S. energy consumption and 21.4% of electricity generation in 2020. Several states, including California, have set targets for increasing the use of renewable energy in their electricity mix.

Energy efficiency: Improving energy efficiency in buildings, appliances, and industrial processes can help reduce greenhouse gas emissions by reducing the amount of energy required to perform the same tasks. The U.S. Department of Energy has implemented several programs to promote energy efficiency, such as the Better Buildings Initiative, which aims to make commercial, industrial, and residential buildings 20% more energy efficient by 2025.

Carbon capture and storage: Carbon capture and storage (CCS) technologies can capture carbon dioxide emissions from industrial processes and power plants and store them underground, preventing them from entering the atmosphere. The U.S. Department of Energy has invested in several CCS projects across the country.

Forest conservation and management: Forests absorb and store large amounts of carbon dioxide, so protecting and managing forests can be an effective way to mitigate climate change. The U.S. Forest Service has implemented several programs to promote forest conservation and sustainable management.

Climate policies at the state level: Several U.S. states, such as California, have implemented their own climate policies to reduce greenhouse gas emissions, such as cap-and-trade programs and renewable portfolio standards.

The Inflation Reduction Act of 2022, which is a major policy that invests in clean energy, infrastructure, innovation, and environmental justice. The act aims to reduce greenhouse gas emissions by 50% by 2030 and achieve net zero emissions by 2050.

The USAID Climate Strategy 2022-2030, which is a plan that guides the U.S. Agency for International Development (USAID) in supporting partner countries to mitigate and adapt to climate change. The strategy focuses on three objectives: accelerating the transition to a low-carbon future, enhancing resilience to climate impacts, and mobilizing climate finance.

The EPA Climate Adaptation Action Plan, which is a document that outlines the steps that the Environmental Protection Agency (EPA) will take to address the impacts of climate change on communities across the nation. The plan covers six areas: protecting human health and the environment, advancing environmental justice and equity, strengthening partnerships and collaboration, enhancing science and data, building capacity and leadership, and mainstreaming adaptation into EPA programs and operations.

Europe's Climate Change Mitigation Strategy:

Europe has been at the forefront of global efforts to address climate change by implementing a range of mitigation strategies aimed at reducing greenhouse gas emissions. One of the most effective strategies that Europe has adopted is the promotion of renewable energy and energy efficiency measures to reduce greenhouse gas emissions.

Renewable Energy

Europe has set ambitious targets for renewable energy deployment, aiming to achieve at least 32% of renewable energy by 2030. Europe has made significant progress towards achieving this target through the deployment of various renewable energy technologies such as wind, solar, and hydropower.

Wind energy has been one of the most successful renewable energy technologies in Europe, with a capacity of over 200 GW. Europe has also made significant strides in the deployment of solar energy, with installed capacity exceeding 140 GW. Hydropower is another important renewable energy source, with Europe having over 200 GW of installed capacity.

To further promote renewable energy deployment, Europe has implemented various policies and incentives such as feed-in tariffs, tax credits, and subsidies. These policies have been instrumental in reducing the cost of renewable energy technologies and incentivizing their adoption.

Energy Efficiency

In addition to promoting renewable energy, Europe has also implemented energy efficiency measures to reduce greenhouse gas emissions. Europe has set a target to reduce energy consumption by at least 32.5% by 2030. Europe has made significant progress towards achieving this target through the implementation of various energy efficiency measures such as building codes, energy labeling, and energy audits.

Building codes have been particularly effective in reducing energy consumption in buildings. Europe has implemented strict building codes that require new buildings to meet certain energy efficiency standards. Europe has also implemented energy labeling schemes that provide consumers with information on the energy efficiency of products such as appliances, lighting, and heating and cooling systems.

Energy audits have also been effective in identifying energy-saving opportunities in buildings and industrial processes. Europe has implemented mandatory energy audits for large companies, which have helped to identify energy-saving opportunities and reduce energy consumption.

Several European countries have been major contributors to the promotion of renewable energy and energy efficiency measures in Europe. The actual contribution of these countries varies depending on their population, economy, and resources.

Germany has been a major contributor to the promotion of renewable energy in Europe. Germany has the largest installed capacity of solar and wind power in Europe, with a total of 129 GW of installed capacity as of 2021. Germany has also been a leader in energy efficiency measures, with strict building codes and energy labeling schemes.

Denmark has been another major contributor to the promotion of renewable energy in Europe. Denmark has the highest share of wind power in its electricity mix, with wind power contributing over 40% of its electricity consumption in 2020. Denmark has also been a leader in energy efficiency measures, with a range of policies and incentives aimed at reducing energy consumption.

Spain has also been a significant contributor to the promotion of renewable energy in Europe. Spain has the second-highest installed capacity of solar power in Europe, with a total of 11 GW of installed capacity as of 2021. Spain has also been a leader in energy efficiency measures, with mandatory energy audits and strict building codes.

France has been a major contributor to the promotion of renewable energy in Europe, with a total installed capacity of 28 GW of renewable energy as of 2021. France has been particularly successful in the deployment of nuclear power, which accounts for over 70% of its electricity generation. France has also implemented energy efficiency measures such as building codes and energy labeling schemes.

Italy has also made significant contributions to the promotion of renewable energy in Europe. Italy has a total installed capacity of 60 GW of renewable energy, with solar power accounting for over 21 GW of installed capacity. Italy has also implemented energy efficiency measures such as building codes, energy audits, and energy labeling schemes.

The United Kingdom has been a leader in offshore wind power, with a total installed capacity of 11 GW of offshore wind power as of 2021. The United Kingdom has also made significant progress in the deployment of other renewable energy technologies such as solar and onshore wind. The United Kingdom has implemented energy efficiency measures such as building codes and energy labeling schemes, and has set ambitious targets for energy efficiency improvements.

Europe's efforts to reduce greenhouse gas emissions through renewable energy and energy efficiency measures have been successful. Europe's ambitious targets for renewable energy deployment and energy efficiency measures have been instrumental in reducing greenhouse gas emissions and combating climate change. Europe's experience provides a useful model for other regions looking to reduce their greenhouse gas emissions and transition to a low-carbon economy.

China's Climate Change Mitigation Strategy

According to the Global Carbon Project, China was the world's largest emitter of greenhouse gases in 2020, accounting for 27% of global carbon dioxide emissions. This is followed by the United States at 11% and India at 6.6%.

According to BBC News, China became the world's largest emitter of carbon dioxide in 2006 and is now responsible for more than a quarter of the world's overall greenhouse gas emissions. In 2021, China was the world's largest methane emitter annually with 58.4 million tonnes of methane emitted. China emits more greenhouse gas than the entire developed world combined. In 2020, China emitted 10,668 million metric tons of carbon dioxide gas, which is the largest amount emitted by any country in the world.

China's high level of emissions is due to its large population, rapid industrialization, and heavy reliance on coal for energy production. In recent years, however, China has taken significant steps to reduce its emissions and transition to a more sustainable economy through the promotion of renewable energy and energy efficiency measures.

China, as the world's largest emitter of greenhouse gases, has been implementing a range of mitigation strategies to address climate change by reducing greenhouse gas emissions through renewable energy and energy efficiency measures.

Renewable Energy

China has made significant progress in the deployment of renewable energy technologies such as wind, solar, and hydropower. China is currently the world's largest producer of solar panels and has the largest installed capacity of wind power in the world.

According to the International Energy Agency (IEA), China was the world's largest producer of solar panels in 2020, accounting for 80% of global solar panel production. In 2020, China's solar panel production capacity was 250 GW, and the country's solar panel exports amounted to $18.7 billion.

China also has the largest installed capacity of wind power in the world. According to the Global Wind Energy Council (GWEC), China's installed wind power capacity was 281 GW at the end of 2020, accounting for 39% of the world's total installed wind power capacity. In 2020, China added 72 GW of new wind power capacity, which was more than the rest of the world combined.

China is also the world's largest producer of hydropower, with an installed capacity of 370 GW in 2020, according to the IEA. China's hydropower generation accounted for 18% of the world's total hydropower generation.

In 2020, China announced its plans to reach peak carbon emissions by 2030 and achieve carbon neutrality by 2060. To achieve these goals, China plans to increase the share of non-fossil fuels in its primary energy consumption to around 25% by 2030, and to achieve 1,200 GW of installed capacity for renewable energy by 2030.

China has set ambitious targets for renewable energy deployment, with a target of achieving 35% of renewable energy in its energy mix by 2030. To achieve this target, China has implemented various policies and incentives such as feed-in tariffs, tax credits, and subsidies to promote renewable energy development. China has also implemented a carbon trading system that provides financial incentives for the deployment of renewable energy.

Energy Efficiency

China has implemented various energy efficiency measures to reduce energy consumption and greenhouse gas emissions. China has set a target to reduce energy consumption per unit of GDP by 15% by 2025. China has made significant progress towards achieving this target through the implementation of various energy efficiency measures such as building codes, energy labeling, and energy audits.

China has implemented strict building codes that require new buildings to meet certain energy efficiency standards. China has also implemented energy labeling schemes that provide consumers with information on the energy efficiency of products such as appliances, lighting, and heating and cooling systems.

Energy audits have also been effective in identifying energy-saving opportunities in buildings and industrial processes. China has implemented mandatory energy audits for large companies, which have helped to identify energy-saving opportunities and reduce energy consumption.

China's efforts to reduce greenhouse gas emissions through renewable energy and energy efficiency measures have been significant. China's ambitious targets for renewable energy deployment and energy efficiency measures have been instrumental in reducing greenhouse gas emissions and combating climate change. China's experience provides a useful model for other regions looking to reduce their greenhouse gas emissions and transition to a low-carbon economy.

India's Climate Change Mitigation Strategy

India, being one of the fastest-growing economies and one of the world's largest energy consumers, is also one of the countries most vulnerable to the impacts of climate change. Therefore, the country has taken significant steps towards mitigating climate change, primarily by reducing greenhouse gas emissions through renewable energy and energy efficiency.

Renewable Energy

India has set ambitious targets for renewable energy deployment to meet the growing energy demand and reduce its dependence on fossil fuels. The country aims to achieve 450 GW of renewable energy capacity by 2030, with 280 GW from solar, 140 GW from wind, 10 GW from bioenergy, and 5 GW from small hydro. The government has also initiated various policy measures and financial incentives to encourage the deployment of renewable energy, such as the National Solar Mission, Renewable Purchase Obligations (RPOs), and the Green Energy Corridor project.

Energy Efficiency

India has also taken various measures to improve energy efficiency across sectors. In 2019, the government launched the Perform, Achieve and Trade (PAT) scheme, which aims to improve energy efficiency in India's large industries. The scheme covers eight energy-intensive sectors, including cement, aluminum, and textiles, and requires industries to meet energy efficiency targets or face penalties. Additionally, the government has launched the UJALA (Unnat Jyoti by Affordable LEDs for All) scheme to promote energy-efficient lighting, and the Standards and Labelling (S&L) program, which mandates energy efficiency labeling for appliances and equipment.

Other Mitigation Strategies

Apart from renewable energy and energy efficiency, India has also implemented other mitigation strategies to address climate change. One such initiative is the National Clean Energy Fund (NCEF), which aims to support research and development of clean energy technologies and provide financial assistance to projects that promote clean energy. The government has also launched the National Electric Mobility Mission Plan (NEMMP) to promote electric vehicles' adoption, reduce emissions from the transportation sector, and decrease the country's dependence on oil imports.

India has taken significant steps towards mitigating climate change by reducing greenhouse gas emissions through renewable energy and energy efficiency measures. While the country has made substantial progress in this direction, much more needs to be done to achieve its ambitious targets and transition to a low-carbon economy. However, with sustained efforts and continued support from the government and private sector, India can move towards a more sustainable and resilient future.

Russia's Climate Change Mitigation Strategy

Russia, being one of the largest producers of fossil fuels, is also a significant emitter of greenhouse gases. However, the country has taken some steps towards reducing greenhouse gas emissions through renewable energy and energy efficiency measures.

Renewable Energy

Russia has set a target of increasing the share of renewable energy in its energy mix to 4.5% by 2024 and to 6% by 2035. To achieve these targets, the government has implemented various policies and measures, such as feed-in tariffs, investment subsidies, and tax incentives. Russia also has significant potential for renewable energy, particularly in hydropower, wind, and solar.

According to the Global Carbon Atlas, Russia was the fourth-largest emitter of carbon dioxide (CO_2) in the world in 2019, accounting for 4.6% of global CO_2 emissions. The country emitted a total of 1.8 billion metric tons of CO_2 in that year.

Moreover, Russia is the world's largest producer of natural gas and the second-largest producer of oil. These fossil fuels are significant sources of greenhouse gas emissions, particularly CO_2 and methane (CH4). Russia's oil and gas industry accounted for approximately 27% of the country's greenhouse gas emissions in 2019, according to the Russian Federal Service for Hydrometeorology and Environmental Monitoring (Roshydromet).

In addition to the oil and gas sector, other sectors that contribute significantly to Russia's greenhouse gas emissions include energy, industry, and transportation. The power sector, for instance, accounted for 29% of the country's greenhouse gas emissions in 2019.

Hydropower is the most widely used renewable energy source in Russia, and the country has the world's third-largest hydropower capacity. Russia has also started developing wind energy, with a target to install 3.3 GW of wind power by 2024. The government has also launched various initiatives to promote solar energy, such as the Solar Energy Development Fund and the Solar Energy Association.

gewo-intl.org/the-threat-to-earth/

Energy Efficiency

Russia has also implemented various measures to improve energy efficiency, particularly in the building and industrial sectors. The government has launched programs such as the Energy Efficiency Standard, which sets energy efficiency standards for buildings, and the Energy Service Company (ESCO) model, which provides financing for energy efficiency projects. Russia has also implemented mandatory energy efficiency labeling for appliances and equipment.

Other Mitigation Strategies

Apart from renewable energy and energy efficiency, Russia has also taken other measures to address climate change. The government has launched initiatives to reduce emissions from the transportation sector, such as promoting the use of electric vehicles and introducing fuel efficiency standards for cars. Russia has also launched several initiatives to reduce methane emissions from the oil and gas industry, which is a significant contributor to the country's greenhouse gas emissions.

While Russia has made some efforts towards mitigating climate change by reducing greenhouse gas emissions through renewable energy and energy efficiency measures, there is still a long way to go. The country's dependence on fossil fuels and its relatively low share of renewable energy in its energy mix, pose significant challenges. However, with sustained efforts and the implementation of more ambitious policies and measures, Russia can move towards a more sustainable and low-carbon future.

Adaptation Measures to Reduce Vulnerability to the Impacts of Climate Change

Climate change is causing significant environmental and socioeconomic impacts worldwide, including rising temperatures, changing precipitation patterns, sea level rise, and more frequent extreme weather events such as hurricanes, droughts, and floods. These impacts are leading to significant risks and vulnerabilities to societies and economies globally, particularly in developing countries and vulnerable communities.

Adaptation measures are strategies and actions taken to reduce vulnerability and enhance resilience to the impacts of climate change. Such measures are designed to help individuals, communities, and countries adapt to changing climatic conditions and protect themselves from the negative effects of climate change. There are several adaptation measures that have been developed to address these impacts, including:

Building Climate Resilient Infrastructure: Infrastructure development, such as roads, bridges, and water supply systems, can be designed to withstand extreme weather events caused by climate change. For example, the city of New York has adopted the New York City Panel on Climate Change (NPCC) recommendations, which include developing more resilient infrastructure and improving coastal flood protection measures.

Implementing Climate-Smart Agriculture: Climate-smart agriculture practices can help farmers to adapt to changing climatic conditions and reduce their vulnerability to crop failures. For example, farmers in Uganda have started using drought-resistant crops and soil conservation techniques to adapt to changing rainfall patterns and reduce their vulnerability to crop failure.

Protecting Natural Ecosystems: Natural ecosystems, such as forests, wetlands, and coral reefs, provide numerous benefits to society, including carbon sequestration and protection against natural disasters. Protecting these ecosystems can help reduce vulnerability to climate change impacts. For example, the restoration of mangrove forests in Vietnam has helped to reduce the impact of storm surges and flooding caused by sea level rise.

Enhancing Disaster Preparedness: Enhancing disaster preparedness can help reduce the impact of extreme weather events caused by climate change. This includes early warning systems, emergency response plans, and evacuation plans. For example, Bangladesh has implemented a comprehensive disaster management program, which includes early warning systems, evacuation plans, and emergency shelters, to reduce the impact of floods and cyclones.

Improving Water Management: Climate change is expected to lead to changes in precipitation patterns and increase the frequency and intensity of extreme weather events, such as floods and droughts. Improving water management can help reduce vulnerability to these impacts. For example, in Ethiopia, the implementation of small-scale irrigation schemes has helped to reduce the vulnerability of farmers to droughts and enhance their food security.

Developing Climate-Resilient Housing: Buildings and homes can be designed to withstand extreme weather events, such as hurricanes and floods, caused by climate change. This can include using building materials that are more resilient to wind and water damage, elevating buildings above flood levels, and designing drainage systems that can handle heavy rainfall. For example, in the Netherlands, the government has implemented policies to encourage the construction of flood-resistant buildings and homes.

Promoting Renewable Energy: The transition to renewable energy sources, such as wind, solar, and geothermal energy, can help reduce greenhouse gas emissions and mitigate the impacts of climate change. This can also provide benefits such as improved air quality and energy security. For example, Costa Rica has set a goal to become carbon neutral by 2050 and has made significant investments in renewable energy, including wind, hydroelectric, and geothermal energy.

Enhancing Public Health Systems: Climate change can have significant impacts on public health, including the spread of vector-borne diseases, such as malaria and dengue fever, and heat-related illnesses. Enhancing public health systems can help reduce vulnerability to these impacts. For example, in Tanzania, the government has implemented a program to distribute mosquito nets to prevent the spread of malaria, which is expected to increase as temperatures rise.

Improving Coastal Management: Climate change is causing sea levels to rise, which can lead to coastal erosion and flooding. Improving coastal management can help reduce vulnerability to these impacts. This can include measures such as beach nourishment, dune restoration, and the creation of wetlands to act as natural barriers against storm surges. For example, in the United States, the state of Louisiana has implemented a comprehensive coastal management plan to protect against coastal erosion and flooding caused by sea level rise.

Encouraging Sustainable Land Use Practices: Sustainable land use practices can help reduce greenhouse gas emissions and enhance the resilience of ecosystems to the impacts of climate change. This can include measures such as sustainable forestry, agroforestry, and conservation agriculture. For example, in Kenya, the government has implemented a program to promote agroforestry practices among smallholder farmers, which has helped to improve soil fertility, increase crop yields, and reduce greenhouse gas emissions.

Strengthening Disaster Risk Reduction: Disaster risk reduction (DRR) strategies can help reduce vulnerability to climate change impacts by minimizing the damage caused by extreme weather events. This can include measures such as early warning systems, emergency preparedness planning, and the development of resilient infrastructure. For example, the government of Japan has implemented a comprehensive disaster risk reduction strategy that includes earthquake and tsunami warning systems, evacuation plans, and the construction of earthquake-resistant buildings.

Encouraging Sustainable Water Management: Climate change can lead to changes in precipitation patterns and water availability, which can have significant impacts on water resources. Sustainable water management practices can help reduce vulnerability to these impacts by improving water use efficiency and enhancing the resilience of water systems. This can include measures such as rainwater harvesting, water conservation, and the development of drought-resistant crops. For example, in India, the government has implemented a program to promote rainwater harvesting and the construction of small-scale water harvesting structures, which has helped to increase water availability in water-scarce areas.

Investing in Climate-Smart Agriculture: Climate-smart agriculture practices can help reduce vulnerability to the impacts of climate change by improving the resilience of agricultural systems to changing climatic conditions. This can include measures such as crop diversification, conservation agriculture, and the use of drought-resistant crops. For example, in Ethiopia, the government has implemented a program to promote climate-smart agriculture practices among smallholder farmers, which has helped to improve soil fertility, increase crop yields, and reduce greenhouse gas emissions.

Implementing Ecosystem-Based Adaptation: Ecosystem-based adaptation (EBA) strategies can help reduce vulnerability to the impacts of climate change by enhancing the resilience of ecosystems and the services they provide. This can include measures such as reforestation, habitat restoration, and the protection of wetlands and coral reefs. For example, in the Philippines, the government has implemented a program to restore mangrove forests, which has helped to improve coastal protection against storm surges and enhance fisheries productivity.

Implementing Green Infrastructure: Green infrastructure refers to the use of natural or semi-natural systems, such as wetlands, forests, and green roofs, to provide ecosystem services and enhance resilience to climate change impacts. This can include measures such as green roofs, green walls, and urban forests. For example, in New York City, the government has implemented a green infrastructure program that includes the construction of green roofs and the planting of trees to reduce stormwater runoff and improve air quality.

Enhancing Climate Information Services: Climate information services can help improve decision-making and reduce vulnerability to climate change impacts by providing accurate and timely information about climate risks and vulnerabilities. This can include measures such as climate modeling, climate forecasting, and the development of early warning systems. For example, in Bangladesh, the government has implemented a program to provide farmers with climate information services, which has helped them to make more informed decisions about crop management and reduce their vulnerability to climate-related risks.

Building Community Resilience: Community resilience refers to the ability of communities to adapt to changing climatic conditions and recover from the impacts of climate change. Building community resilience can include measures such as community-based disaster risk reduction, capacity building, and social safety nets. For example, in Nepal, the government has implemented a program to build the resilience of local communities to the impacts of climate change, which has helped to reduce their vulnerability to climate-related risks and enhance their capacity to adapt to changing climatic conditions.

Strengthening Climate Governance: Effective climate governance is essential for reducing vulnerability to climate change impacts by ensuring that policies and actions are aligned with climate objectives and that resources are effectively allocated to adaptation measures. This can include measures such as the development of national adaptation plans, the implementation of climate policies, and the establishment of climate finance mechanisms. For example, in the Maldives, the government has implemented a comprehensive climate governance framework that includes the development of a national adaptation plan and the establishment of a climate change trust fund.

Promoting Sustainable Urban Development: As urbanization continues to grow around the world, it is essential to promote sustainable urban development practices that can reduce vulnerability to climate change impacts. This can include measures such as green building design, sustainable transportation, and the development of green spaces. For example, in Singapore, the government has implemented a comprehensive sustainable urban development program that includes the construction of green buildings and the development of sustainable transportation systems.

Enhancing Health Resilience: Climate change can have significant impacts on human health, particularly through the spread of vector-borne diseases, heat stress, and food and waterborne illnesses. Enhancing health resilience can include measures such as strengthening health systems, improving disease surveillance, and promoting healthy lifestyles. For example, in Uganda, the government has implemented a program to strengthen health systems and enhance disease surveillance in response to the impacts of climate change.

Promoting Climate-Sensitive Tourism: Tourism is an important sector for many countries, but it can also be vulnerable to the impacts of climate change, particularly through changes in temperature and precipitation patterns. Promoting climate-sensitive tourism practices can include measures such as promoting sustainable tourism, diversifying tourism activities, and developing climate adaptation strategies. For example, in the Caribbean, several countries have implemented climate-sensitive tourism programs that include the development of sustainable tourism practices and the promotion of climate-resilient tourism activities.

In conclusion, there are numerous adaptation measures that can be taken to reduce vulnerability to the impacts of climate change. These measures are essential in helping communities and countries to adapt to changing climatic conditions and protect themselves from the negative effects of climate change. By implementing these measures, we can build more resilient societies and economies that are better prepared to face the challenges of climate change.

Importance of International Cooperation and Policy Action

International cooperation and policy action on climate change are critical for addressing the global challenge of climate change. Climate change is a global problem that requires global solutions, and international cooperation and policy action can play a vital role in addressing this issue. Here are some reasons why international cooperation and policy action are essential:

Climate change is a global problem: Climate change affects everyone, regardless of where they live or what they do. It is a global problem that requires a coordinated response from all countries. International cooperation and policy action can help to bring countries together to address this challenge.

Climate change requires collective action: Addressing climate change requires collective action from all countries. No single country can solve the problem alone. International cooperation and policy action can help to facilitate collective action and ensure that all countries are doing their part to reduce greenhouse gas emissions and mitigate the impacts of climate change.

International cooperation can lead to more effective policies: International cooperation can help to identify best practices and share knowledge on effective policies for addressing climate change. This can lead to the development of more effective policies that can help to reduce greenhouse gas emissions and increase resilience to climate change impacts.

International cooperation can mobilize resources: Addressing climate change requires significant resources, including funding, technology, and expertise. International cooperation can help to mobilize these resources and ensure that they are used effectively to support climate action.

International cooperation can promote equity: Climate change disproportionately affects vulnerable populations, including low-income communities and indigenous peoples. International cooperation can help to promote equity and ensure that these communities are not left behind in the transition to a low-carbon economy.

There are many examples of international cooperation and policy action on climate change that have been successful in addressing this global challenge. For example:

The Paris Agreement: In 2015, 197 countries signed the Paris Agreement, which sets a target to limit global warming to well below 2°C above pre-industrial levels and pursue efforts to limit the temperature increase to 1.5°C. The Paris Agreement also includes provisions for countries to regularly report on their emissions and progress towards meeting their targets.

The Montreal Protocol: The Montreal Protocol is an international treaty that was signed in 1987 to phase out the production and consumption of ozone-depleting substances. The treaty has been successful in reducing the use of these substances and has helped to protect the Earth's ozone layer.

The Clean Development Mechanism: The Clean Development Mechanism (CDM) is a mechanism under the United Nations Framework Convention on Climate Change (UNFCCC) that allows developed countries to offset their greenhouse gas emissions by investing in emission reduction projects in developing countries. The CDM has been successful in promoting clean energy projects in developing countries and helping to reduce greenhouse gas emissions.

The Global Alliance for Buildings and Construction: The Global Alliance for Buildings and Construction is a global initiative launched in 2015 to promote the decarbonization of the building sector. The initiative brings together governments, cities, private sector organizations, and civil society to promote the adoption of low-carbon building practices and technologies.

The Renewable Energy Policy Network for the 21st Century: The Renewable Energy Policy Network for the 21st Century (REN21) is a global network of governments, research organizations, and civil society organizations working to promote the transition to renewable energy. REN21 produces annual reports on the status of renewable energy around the world and provides guidance to governments and other stakeholders on how to accelerate the transition to renewable energy.

The Global Environment Facility: The Global Environment Facility (GEF) is a financial mechanism established in 1991 to help countries address global environmental issues, including climate change. The GEF provides funding to support climate mitigation and adaptation projects in developing countries.

The Climate and Clean Air Coalition: The Climate and Clean Air Coalition (CCAC) is a global initiative launched in 2012 to reduce short-lived climate pollutants, such as black carbon and methane. The CCAC brings together governments, civil society organizations, and private sector companies to promote the adoption of measures to reduce these pollutants, which have significant impacts on climate change and public health.

The International Solar Alliance: The International Solar Alliance is a global initiative launched in 2015 to promote the deployment of solar energy around the world. The initiative brings together countries with high solar potential to promote the adoption of solar energy and to develop policies and strategies to support its deployment.

In conclusion, international cooperation and policy action on climate change are essential for addressing this global challenge. Climate change requires collective action, effective policies, and significant resources to address, and international cooperation can help to facilitate this. These initiatives are just a few examples of the many international cooperation and policy action efforts underway to address climate change. By working together, countries, organizations, and individuals can help to address this critical issue and build a more sustainable future for all.

The Net Zero Vision

The Net Zero Vision is a global initiative aimed at achieving a carbon-neutral world by balancing greenhouse gas emissions with their removal from the atmosphere. This means that the amount of greenhouse gases released into the atmosphere, mainly carbon dioxide, should be equal to the amount removed from the atmosphere through natural or artificial means. The vision seeks to limit global warming to a maximum of 1.5°C above pre-industrial levels, in line with the Paris Agreement on Climate Change.

The Net Zero Vision is gaining momentum globally, with countries, cities, and corporations committing to achieving net-zero emissions by 2050 or earlier. The Intergovernmental Panel on Climate Change (IPCC) report, released in 2021, warned that the world must achieve net-zero emissions by 2050 to avoid catastrophic climate change.

One of the key drivers of the Net Zero Vision is renewable energy. The falling costs of renewable energy technologies such as solar, wind, and hydropower are making it more feasible for countries and corporations to transition to clean energy. According to a report by the International

Energy Agency, renewables accounted for 72% of global power capacity additions in 2019.

In addition to renewable energy, other sectors are also making significant progress in decarbonization. For example, the transportation sector is moving towards electric vehicles, with major car manufacturers such as Tesla, General Motors, and Volkswagen investing heavily in EV technology. The aviation sector is also exploring the use of biofuels and hydrogen as alternatives to fossil fuels.

Governments are also playing a crucial role in driving the Net Zero Vision. For example, the UK has set a target of achieving net-zero emissions by 2050 and has put in place policies to support the transition to clean energy, such as subsidies for electric vehicles and renewable energy projects. The EU has set a similar target and is investing heavily in renewable energy and energy efficiency measures.

The Net Zero Vision is not only important for mitigating the effects of climate change but also presents economic opportunities. According to a report by the International Renewable Energy Agency, the transition to renewable energy could create over 40 million jobs globally by 2050.

There are, however, challenges to achieving the Net Zero Vision. For example, some sectors such as heavy industry and agriculture are more difficult to decarbonize than others. Additionally, there is a need for significant investment in clean energy infrastructure and technologies, which could be a barrier for some countries and corporations.

Important Aspects of Net Zero Vision

Role of carbon capture and storage (CCS): CCS technologies involve capturing carbon dioxide from industrial processes and storing it underground or using it for enhanced oil recovery. CCS can play a crucial role in achieving the Net Zero Vision, particularly for sectors that are difficult to decarbonize, such as heavy industry and aviation. According to the International Energy Agency, CCS could account for nearly 15% of global emissions reductions by 2050. However, there are challenges to the widespread deployment of CCS, such as high costs, technical barriers, and concerns about the long-term safety of carbon storage.

Potential impact on developing countries: The Net Zero Vision presents both opportunities and challenges for developing countries. On the one hand, transitioning to renewable energy and decarbonizing key sectors could bring economic benefits and help address issues such as energy poverty and air pollution. On the other hand, some developing countries may struggle to access the necessary funding and technology to support the transition to clean energy, and may also face the risk of carbon leakage, where emissions-intensive industries relocate to countries with less stringent climate policies.

Importance of public engagement and awareness: Public engagement and awareness are critical for achieving the Net Zero Vision, as the transition to a carbon-neutral world will require significant changes in behavior and consumption patterns. Education, outreach, and communication efforts can help build support for climate action and encourage individuals, communities, and businesses to take steps to reduce their carbon footprint. Engaging with stakeholders and ensuring that their voices are heard can also help build trust and support for climate policies and initiatives.

Initiatives, policies, and organizations dedicated to advancing the Net Zero Vision: There are many global initiatives, policies, and organizations dedicated to advancing the Net Zero Vision. These include:

- The UN's Race to Zero campaign, which aims to rally leadership and support from businesses, cities, regions, and investors to achieve net-zero emissions by 2050 or earlier.
- The Net Zero Asset Owners Alliance, a group of investors representing over $6.6 trillion in assets under management who have

committed to transitioning their portfolios to net-zero emissions by 2050.

- The International Energy Agency's Net Zero by 2050 roadmap, which provides a detailed blueprint for achieving a net-zero emissions global energy system by 2050, including recommendations for policy and investment decisions.
- The Paris Agreement on Climate Change, which was signed by 197 countries and sets a target of limiting global warming to well below 2°C above pre-industrial levels and pursuing efforts to limit warming to 1.5°C.

Overall, the Net Zero Vision is a complex and multifaceted initiative that requires collaboration, innovation, and investment from governments, businesses, and individuals around the world. While there are many challenges and uncertainties, there is also significant potential for economic growth, job creation, and improved health and well-being as we transition to a more sustainable and equitable future.

Constraints That Could Limit Progress Towards the Net Zero Vision

Infrastructure challenges: Achieving a net-zero emissions economy will require significant upgrades to existing infrastructure, including transportation networks, buildings, and energy systems. This may require large-scale investments and a significant amount of time to implement, which could delay progress towards the Net Zero Vision.

Carbon-intensive industries: Certain industries, such as aviation, shipping, and heavy manufacturing, have very high carbon footprints and may be difficult to decarbonize. It may be challenging to develop low-carbon alternatives for these industries, and transitioning away from carbon-intensive processes could be expensive and require significant technological advances.

gewo-intl.org/the-threat-to-earth/

Limited public support: While there is growing public concern about climate change, there may be limited support for some of the policies and measures needed to achieve the Net Zero Vision. For example, some people may be resistant to higher taxes or fees on carbon-intensive activities, or to changes in their daily habits that may be necessary to reduce emissions.

Limited availability of sustainable materials: As the demand for sustainable products increases, there may be limited availability of key materials, such as rare earth metals, which are used in many renewable energy technologies. This could lead to supply chain disruptions and higher costs for these materials.

Natural disasters and climate-related events: Climate change is leading to more frequent and severe weather events, such as hurricanes, floods, and wildfires, which can damage infrastructure and disrupt supply chains. These events can be costly to repair and may delay progress towards the Net Zero Vision.

Overall, while there is significant momentum towards achieving the Net Zero Vision, there are still many challenges and constraints that could limit progress. Overcoming these challenges will require innovative solutions, strong leadership, and a commitment from all stakeholders to work towards a sustainable and equitable future.

Bill Gates' View on Net Zero Vision

Bill Gates has been a vocal advocate for the Net Zero Vision and has emphasized the urgent need for action to address climate change. In his book, *How to Avoid a Climate Disaster*, Gates lays out his view on the importance of achieving net-zero emissions and provides recommendations on how to do so.

Gates argues that achieving net-zero emissions will require a global effort to reduce greenhouse gas emissions across all sectors of the economy. He suggests that the most important steps towards achieving this goal are:

Innovating and investing in new technologies: Gates emphasizes the need for significant investment in the research, development, and deployment of new technologies that can reduce greenhouse gas emissions. He suggests that governments and private investors should invest in a range of technologies, including renewable energy, nuclear energy, and carbon capture and storage.

Transitioning to renewable energy: Gates suggests that a rapid transition to renewable energy is critical to achieving net-zero emissions. He recommends increasing investment in solar and wind energy, as well as investing in new technologies such as advanced nuclear reactors and energy storage systems.

Transforming the transportation sector: The transportation sector is a major contributor to greenhouse gas emissions, and Gates suggests that electrification of transportation is critical to achieving net-zero emissions. He recommends transitioning to electric cars, buses, and trucks, as well as investing in public transportation systems.

Developing sustainable agriculture and land use practices: Agriculture and land use are significant sources of greenhouse gas emissions, and Gates suggests that sustainable agricultural practices and reforestation efforts can help to reduce emissions.

Encouraging public engagement and action: Gates emphasizes the importance of public awareness and engagement on climate issues. He suggests that individuals can make a difference by taking steps such as reducing their energy consumption and supporting political leaders who prioritize climate action.

gewo-intl.org/the-threat-to-earth/

Overall, Gates believes that achieving net-zero emissions will require a concerted effort from governments, businesses, and individuals around the world. He suggests that a combination of policy measures, technological innovation, and public engagement will be necessary to achieve this goal.

In conclusion, the Net Zero Vision is a global initiative aimed at achieving a carbon-neutral world and limiting global warming to a maximum of 1.5°C above pre-industrial levels. The initiative is gaining momentum globally, with countries, cities, and corporations committing to achieving net-zero emissions by 2050 or earlier. Renewable energy is a key driver of the Net Zero Vision, with other sectors also making significant progress in decarbonization. Governments are also playing a crucial role in driving the transition to clean energy. While there are challenges to achieving the Net Zero Vision, it presents economic opportunities and is crucial for mitigating the effects of climate change.

Chapter 8

Conclusion

Summary of the Impact of Climate Change on Humanity and the Planet

For more than two centuries now, human societies have tapped into a one-time gift from Earth (fossil fuels) to make technological advances that have benefited and advanced society in incredible ways. Fossil fuels in the form of coal, oil, and natural gas have powered new machines, been the basis for remarkable new materials, facilitated vastly more productive agriculture, and led to a technological information revolution that now connects humanity in ways unimaginable even a few short decades ago.

Advancements in human ingenuity and the harnessing of dense economical energy from fossil fuels has allowed for exponential growth of human societies on Earth. Now, however, with a deep understanding of the carbon cycle and how human actions impact climate, we must reduce and eventually eliminate greenhouse gas emissions from fossil fuels (and most other sources) to ensure the conditions we all enjoy on Earth are here for generations to come. In addition, we must work together to adapt to the changing climate we are experiencing because of historical emissions. We must also plan to become more resilient and adapt to climate changes still to come.

Climate change is having significant impacts on both humanity and the planet, with consequences that are both immediate and long term. The Earth's climate has always been subject to change, but the current changes are happening much more rapidly and are largely driven by human activities such as burning fossil fuels and deforestation.

One of the most visible effects of climate change is rising sea levels. As temperatures increase, glaciers and ice sheets melt, leading to a rise in ocean levels. This poses a significant threat to coastal communities, particularly those in low-lying areas, which are at risk of flooding and storm surges. Additionally, changes in precipitation patterns can lead to droughts, which can reduce the availability of freshwater and negatively impact agriculture, food security, and public health.

Climate change is also causing more frequent and severe weather events such as heatwaves, hurricanes, and cyclones. These events can have devastating impacts on communities, causing loss of life and damage to infrastructure. In addition, climate change can exacerbate existing health problems, increase the spread of diseases, and create new ones. For example, the warming of water bodies can lead to the spread of harmful algae blooms, while changes in rainfall patterns can create conditions that are conducive to the spread of diseases like malaria.

The impacts of climate change are felt most acutely by vulnerable populations, including low-income communities, indigenous peoples, and those living in coastal areas. These groups often lack the resources necessary to adapt to the changes that are occurring and are disproportionately affected by the impacts of climate change.

Finally, it is worth noting that the long-term impacts of climate change are still largely unknown. However, it is clear that immediate and sustained action is necessary to mitigate its effects and prevent further damage to both humanity and the planet. This will require global cooperation, significant changes in our energy systems and consumption patterns, and investment in research and development of new technologies that can help us transition to a low-carbon economy.

The Urgent Need for Action to Mitigate and Adapt to the Impacts of Climate Change

The effects of climate change are already being felt across the globe, and the consequences are likely to worsen unless urgent action is taken to mitigate and adapt to its impacts.

One of the most immediate effects of climate change is the increase in global temperatures. The Intergovernmental Panel on Climate Change (IPCC) has warned that the global temperature should not rise more than 1.5 degrees Celsius above pre-industrial levels to avoid the worst impacts of climate change. However, the current trajectory of emissions will lead to a much higher temperature rise, causing severe impacts such as more frequent and intense heatwaves, droughts, and flooding.

The impacts of climate change are not limited to the environment but also have significant social and economic consequences. The increased frequency of extreme weather events can damage infrastructure, disrupt food supplies, and cause displacement and migration. Climate change can also exacerbate existing inequalities, as vulnerable populations, such as the poor, children, and the elderly, are disproportionately affected.

Mitigating climate change requires reducing GHG emissions. This can be achieved through a combination of measures, including increasing energy efficiency, transitioning to renewable energy sources, and improving public transportation. Governments, businesses, and individuals all have a role to play in reducing emissions.

However, even with aggressive mitigation measures, some level of climate change is already locked in due to past emissions. Therefore, adaptation measures are also necessary to minimize the impacts of climate change. These measures include building resilient infrastructure, improving water management, and developing early warning systems for extreme weather events.

The urgency of action to mitigate and adapt to climate change cannot be overstated. Delaying action will only make the problem more challenging and expensive to address. In 2015, 195 countries signed the Paris Agreement, committing to limiting global warming to well below 2 degrees Celsius above pre-industrial levels. However, current commitments are insufficient to achieve this goal.

Therefore, a renewed and more ambitious effort is needed from all sectors to limit global warming to a level that is safe for people and the planet. The urgency of action requires global collaboration, innovation, and bold leadership to tackle the problem. The time for action is now, and we cannot afford to wait any longer to address the urgent need for action to mitigate and adapt to the impacts of climate change.

The CityGreen Blueprint – A London, Ontario Initiative

As I was on the hunt for a property in 2022, my realtor suggested a stunning home in the charming city of London, located in Ontario, Canada. I was ecstatic to inspect the property as I had never been to London before and was eager to explore what the city had to offer. After a 2-hour and 20-minute drive from East Toronto, I finally arrived in London and was immediately struck by the peaceful and organized vibe of the city. It was evident that London, Ontario was a city that took pride in its beauty and sustainability.

Curious to learn more about the city's sustainability efforts, I did some research and was blown away by their Climate Emergency Action Plan. This is undoubtedly one of the best plans in Canada for reducing greenhouse gas emissions and adapting to the impacts of climate change.

gewo-intl.org/the-threat-to-earth/

It's no wonder that the city has surpassed both federal and provincial targets! The CityGreen blueprint is a must-see for any city looking to achieve net-zero community greenhouse gas emissions within a set timeframe.

Needless to say, I fell in love with the city even more and, without hesitation, purchased the property. If you're looking for a city that values sustainability, peace, and organization, then look no further than London, Ontario—the hidden gem of Canada.

The CityGreen Initiative

The City of London, located in Ontario, Canada, has implemented a series of initiatives to address the climate emergency that is affecting the city and its citizens. One of the major initiatives is the CityGreen plan, which aims to make London a more sustainable and resilient city by reducing greenhouse gas emissions and adapting to the impacts of climate change.

The CityGreen plan was developed in consultation with a broad range of stakeholders, including community groups, business leaders, and residents. It outlines specific strategies and actions that the city will take to address climate change, including reducing energy consumption, increasing the use of renewable energy sources, promoting sustainable transportation, and protecting natural areas.

One of the key elements of the CityGreen plan is the Climate Emergency Action Plan (CEAP), which was introduced in 2019 and outlines specific targets and actions to be taken to reduce greenhouse gas emissions in the city, increase resilience to climate change impacts, and engage the community in the transition to a low-carbon future. The plan includes specific targets and actions to achieve the following goals:

- Net-zero emissions by 2050

- 30% reduction in greenhouse gas emissions by 2030
- 100% renewable energy in municipal operations by 2050
- Increase tree canopy cover to 34%
- Increase active transportation mode share to 8% by 2030

To achieve these goals, the plan includes strategies to reduce emissions from transportation, buildings, waste, and energy use. It also includes measures to increase the use of public transit, cycling, and walking, promote energy efficiency in buildings, and support the transition to renewable energy.

The CEAP includes a range of initiatives, such as:

Building Retrofits: The city plans to retrofit municipal buildings to improve energy efficiency, reduce greenhouse gas emissions, and save on energy costs.

Sustainable Transportation: The city plans to increase the use of sustainable transportation options, such as cycling and public transit, to reduce greenhouse gas emissions from transportation.

Renewable Energy: The city plans to increase the use of renewable energy sources, such as solar and wind power, to reduce greenhouse gas emissions from electricity generation.

Waste Reduction: The city plans to reduce the amount of waste sent to landfills by promoting recycling and composting.

To implement the CEAP, the city has developed a series of work plans that outline specific actions and timelines for each initiative. The work plans are regularly reviewed and updated to ensure that the city is making progress towards its targets.

In addition to the actions taken by the city, households are also required to play a role in reducing greenhouse gas emissions. The city has launched a number of programs and initiatives to encourage households to reduce their energy consumption and adopt more sustainable practices, such as:

Energy Efficiency Incentives: The city offers incentives and rebates for households that upgrade to more energy-efficient appliances and systems.

Green Building Standards: The city has adopted green building standards that encourage households to build or renovate their homes to be more energy efficient.

Sustainable Transportation: The city encourages households to use sustainable transportation options, such as cycling, walking, and public transit, to reduce greenhouse gas emissions from transportation.

Waste Reduction: The city encourages households to reduce the amount of waste they produce by composting, recycling, and reducing consumption.

How the CEAP Engages the Household to Take Action

The Climate Emergency Action Plan also features 11 examples of different kinds of households and what they can do to reduce their greenhouse gas emissions. In this example, an average-income household of four, in an older, single-family house in an established neighborhood with two vehicles (one large, one compact), currently produces 3.4 tonnes of greenhouse gas per person.

However, if this household took the following actions in the coming years…

- 20% reduction in heat loss with air sealing and insulation.
- First vehicle changed to hybrid and 10% reduction in kilometers driven.

- Second vehicle replaced with bike/cargo bike/e-bike.
- Reduction in organic waste.
- Basement flooding measures incorporated.
- Raingardens installed.

... they could make their home more resilient and reduce their greenhouse gas emissions down to 1.9 tonnes per person by 2030. That's a 44% reduction!

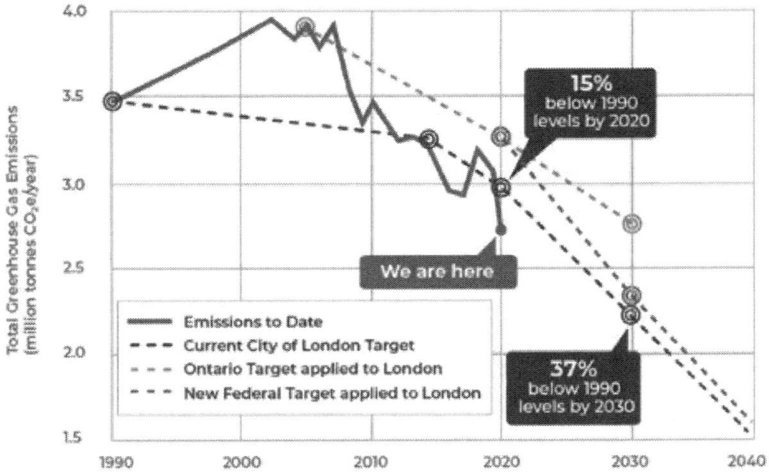

Source: City of London

London's Greenhouse Gas Emissions
Versus Federal and Provincial Reduction Targets

In summary, the CityGreen and Climate Emergency Action Plan initiated by the City of London, Ontario, Canada, aims to reduce greenhouse gas emissions and adapt to the impacts of climate change. The plan outlines specific strategies and actions to be taken by the city and households to achieve these goals, and it includes work plans to guide implementation. The plan is regularly reviewed and updated to ensure progress towards targets, and households are encouraged to adopt more sustainable practices through a range of programs and incentives.

The Carbon Footprint Pledge – Individual Commitment

Individual Commitment Towards Carbon Emission Reduction

- Use only recycled paper.
  Why? Because it saves the atmosphere about five pounds of carbon dioxide per ream of paper.

- Set your thermostat just two degrees warmer in the summer than you're accustomed to and two degrees cooler in the winter.
  Why? Because those tiny adjustments will keep approximately two thousand pounds of carbon dioxide out of the atmosphere per year.

- Don't run your dishwasher until it's completely full.
  Why? Because it will save the atmosphere about one hundred pounds of carbon dioxide per year.

- Pick the three lamps you use most in your house and change the bulbs to easily accessible compact fluorescent bulbs.
  Why? Because you'll eliminate an extra three hundred pounds of carbon dioxide from the atmosphere every year.

- Adjust the heat on your water heater to a maximum of 120 degrees, and have it insulated.
  Why? Because it saves 1,550 pounds of carbon dioxide from the air every year.

- Reduce the duration of your showers by two to three minutes.
  Why? Because 350 pounds of carbon dioxide will be saved every year thanks to less water needing to be heated.

- Check your tires every month to make sure they're property inflated.
  Why? Because it will save 250 pounds of carbon dioxide per year.

- Change your heating and air-conditioning filters, or clean them, as often as recommended.
  Why? Because it will keep the units from having to work harder than they're designed to work to keep you comfortable, as well as keeping an extra 350 pounds of carbon dioxide out of the atmosphere per year.

- Turn your computer off rather than letting it "sleep," and unplug electronics when you're not using them.
  Why? Because it will save a minimum of twelve hundred pounds of carbon dioxide per year.

- Take care of the simple chore of caulking and weather-stripping your exterior doors and windows.
  Why? Because not only will you be more comfortable all year long, but you'll also eliminate seventeen hundred pounds of carbon dioxide per year.

- Plant a tree, or have one planted by an organization like Global Environmental Watch Organization (GEWO-INTL.ORG), in honor of a lost loved one.
  Why? Because you'll be adding more oxygen to the atmosphere and saving an amazing two thousand pounds of carbon dioxide.

As you replace old appliances, buy new ones that bear an "Energy Star" label, designed to save both carbon dioxide emissions and money.

gewo-intl.org/the-threat-to-earth/

Making sure the walls and ceilings of your home are well-insulated can save a minimum of two thousand pounds of carbon dioxide, not to mention hundreds of dollars every year.

Changing single-pane windows to double-pane windows conserves energy, saves a fortune in power bills, and eliminates an amazing ten thousand pounds of carbon dioxide.

Switching from your current showerhead to a low-flow showerhead will save approximately 350 pounds of carbon dioxide and an average of around two hundred dollars per year.

When it's time for a new car, remember that a hybrid will save almost seventeen thousand pounds of carbon dioxide and $3,800 per year. Even a more fuel-efficient car will save thousands and thousands of pounds of carbon dioxide and thousands of dollars at the same time. The average passenger gasoline vehicle emits about 4.6 metric tons of CO_2 per year. If an electric vehicle were driven the same distance as a typical gasoline vehicle, it could save about 2.3 metric tons of CO_2 per year.

Consider switching to home solar energy as this can significantly reduce carbon emissions from electricity consumption in a home. On average, a typical home in the United States emits about 7.5 metric tons of carbon dioxide (CO_2) per year from electricity consumption. Switching to solar energy can reduce these carbon emissions by an estimated 3-4 metric tons per year. Sign up for clean green power from your local provider and enjoy some of the government grants that come with it.

Avoid food wastage. If 50% of food waste is reduced and we adopt a plant-based diet by 2050, 26.2 gigatons of greenhouse gas emissions could be avoided.

Carefully plan your meals, avoid unnecessary purchases, eat leftovers, and don't be afraid of the "ugly" fruits and vegetables (they're usually the best).

Avoid unnecessary air travel. A round-trip flight from New York to Los Angeles, which is 4,938 miles (7,950 kilometers), emits around 1.2 metric tons of carbon dioxide per person. If you must travel, consider flying with airline carriers that use biofuels. Biofuels in airplanes produce 60% less carbon emissions than regular jet fuel (on a life cycle basis).

About 162.6 billion packages are the estimated average amount of boxes and letters delivered around the world each year by FedEx, UPS, and the USPS.

According to a study by the World Economic Forum, the global carbon emissions from parcel deliveries in 2019 were estimated to be around 617 million metric tons.

You can help by bundling your packages so that several items can come in one delivery. Try to avoid ordering next day, as you make it harder for delivery services to organize efficient deliveries.

The Author's Thoughts

As I typed the last word of this book on my laptop, I felt a sense of relief. It was as if a heavy burden had been lifted from my shoulders. I have carried this burden for many years since I was 15. I have always known that climate change is real and that humans are on a path of self-destruction. Like little children who act without considering the consequences of their actions, humans have treated the Earth with greed and without regard for its sustainability.

In this book, you read about my three remarkable encounters with the God of creation in my dreams. These encounters added an extra layer of burden to my heart, as so many questions went through my mind. What if it was just a dream and a figment of my imagination? What if the dream was a sign, warning humans to take action now to save the Earth, or risk losing humanity? What if I do nothing and one day face my creator and must answer for this? What if I do nothing and fade into oblivion, and no one and no God is there to question me on this?

I decided the best approach to these "what if" scenarios was to leave no regrets on my deathbed. If the impending danger posed by climate change is real, I want to do something to stop it or at least prevent it from happening, through my current actions. For this reason, I founded the Global Environmental Watch Organization (GEWO), a nonprofit dedicated to saving planet Earth from the impending threat of climate change. I have also written this book, *The Threat to Earth*, to enlighten humanity about the impending threat of climate change and the necessary actions we must take to prevent it.

Now, my question to you is this: Would you rather join those who deny the reality of climate change, or would you join me in the fight to save the Earth? The biggest regret people have on their deathbed is not the things they did but the things they didn't do. This sentiment could be rephrased as "The greatest tragedy of life is not the mistakes we make but the opportunities we miss by not taking action."

This book is a call to action for all of us. It is time to take responsibility for the future of our planet and act now to save the Earth. The clock is ticking, and the stakes couldn't be higher. Will we rise to the challenge, or will we face the unthinkable consequences of our inaction? The choice is ours. *Act now to save the Earth or risk losing humanity forever!*

Visit ***gewo-intl.org*** to join the global fight against climate change.

Reference

Chapter 1 Introduction of the Threat to Earth

The Earth

NASA: https://www.nasa.gov/audience/forstudents/k-4/stories/nasa-knows/what-is-earth-k4.html

National Geographic: https://www.nationalgeographic.com/science/earth/earths-atmosphere/

Encyclopedia Britannica: https://www.britannica.com/place/Earth

Smithsonian National Museum of Natural History: https://naturalhistory.si.edu/education/teaching-resources/earth-science

The United States Holocaust Memorial Museum: https://encyclopedia.ushmm.org/content/en/article/atomic-bombings-of-hiroshima-and-nagasaki

History.com: https://www.history.com/topics/world-war-ii/bombing-of-hiroshima-and-nagasaki

The National WWII Museum: https://www.nationalww2museum.org/war/topics/atomic-bombing-hiroshima-and-nagasaki

The Hiroshima Peace Memorial Museum: https://www.pcf.city.hiroshima.jp/top_e.html

The Holy Bible: New King James Version. (1982). Genesis 1

Greenhouse Effect: https://climatekids.nasa.gov/greenhouse-effect/)

Greenhouse Gases: https://byjus.com/biology/greenhouse-effect-gases/

Intergovernmental Panel on Climate Change (IPCC). (2018). Global Warming of 1.5°C. IPCC Special Report. https://www.ipcc.ch/sr15/

NASA Global Climate Change. (2021). Climate Change: Vital Signs of the Planet. https://climate.nasa.gov/

United Nations Framework Convention on Climate Change (UNFCCC). (2015). Paris Agreement. https://unfccc.int/process-and-meetings/the-paris-agreement/the-paris-agreement

Chapter 2 Climate Change and Its Effects on the Natural World

"How Does Climate Change Affect Wildlife and Ecosystems?" National Wildlife Federation, https://www.nwf.org/Educational-Resources/Wildlife-Guide/Threats-to-Wildlife/Climate-Change

"Climate Change and Biodiversity Loss," World Wildlife Fund, https://www.worldwildlife.org/threats/climate-change-and-biodiversity-loss

"Climate Change and Its Impact on Species," Conservation International, https://www.conservation.org/stories/climate-change-and-its-impact-on-species

USDA. (2020). Climate Change and Agriculture in the United States: Effects and Adaptation. United States Department of Agriculture. https://www.usda.gov/topics/climate/climate-change-and-agriculture

gewo-intl.org/the-threat-to-earth/

FAO. (2021). The Impact of Disasters and Crises on Agriculture and Food Security. Food and Agriculture Organization of the United Nations. http://www.fao.org/3/cb1912en/cb1912en.pdf

WMO. (2020). State of the Global Climate 2020. World Meteorological Organization. https://public.wmo.int/en/resources/library/state-of-global-climate-2020

"Ocean Acidification: A National Strategy to Meet the Challenges of a Changing Ocean," a report by the National Research Council of the National Academies: https://www.nap.edu/catalog/12904/ocean-acidification-a-national-strategy-to-meet-the-challenges-of

"Ocean Acidification," a resource from the National Oceanic and Atmospheric Administration (NOAA): https://oceanacidification.noaa.gov/

"Impacts of Ocean Acidification on Marine Organisms: Quantifying Sensitivity and Adaptive Capacity," a paper by Philip L. Munday et al. published in Reviews in Aquaculture: https://onlinelibrary.wiley.com/doi/full/10.1111/raq.12104

"Ocean Acidification and Its Effects on Marine Ecosystems," a paper by Ulf Riebesell et al. published in Annals of the New York Academy of Sciences: https://nyaspubs.onlinelibrary.wiley.com/doi/full/10.1111/j.1749-6632.2009.04164.x

"Ocean Acidification: The Other CO_2 Problem," an article by Richard A. Feely and Christopher L. Sabine published in Annual Review of Marine Science: https://www.annualreviews.org/doi/abs/10.1146/annurev.marine.010908.163834

Chapter 3 Climate Change and Its Impact on Human Health

"Climate Change and Health: Air Pollution," a resource from the World Health Organization (WHO): https://www.who.int/globalchange/climate/summary/en/index5.html

"Air Pollution and Climate Change," a resource from the Environmental Protection Agency (EPA): https://www.epa.gov/air-pollution-transportation/air-pollution-and-climate-change

"Climate Change and Air Pollution: Effects on Public Health," a paper by Lisa Patel et al. published in the American Journal of Public Health: https://www.ncbi.nlm.nih.gov/pmc/articles/PMC7104984/

"Wildfires, Global Climate Change, and Human Health," a paper by Irva Hertz-Picciotto and Kent E. Pinkerton published in New England Journal of Medicine: https://www.nejm.org/doi/full/10.1056/NEJMp0807571

"The Impacts of Climate Change on Human Health in the United States: A Scientific Assessment," a report by the U.S. Global Change Research Program: https://health2016.globalchange.gov/

"Climate Change and Health: Insect-Borne Diseases," a resource from the World Health Organization (WHO): https://www.who.int/globalchange/climate/summary/en/index7.html

"Climate Change and Vector-Borne Diseases," a resource from the Centers for Disease Control and Prevention (CDC): https://www.cdc.gov/climateandhealth/effects/vector-borne-diseases.html

"Infectious Diseases, Climate Change, and Globalization," a paper by Jonathan A. Patz et al. published in the New England Journal of Medicine: https://www.nejm.org/doi/full/10.1056/NEJMra020201

gewo-intl.org/the-threat-to-earth/

"Climate Change and the Spread of Vector-Borne Diseases: Using Knowledge of Mechanisms to Mitigate Impact," a paper by Sara E. Randolph and Matthew B. Thomas published in Environmental Health Perspectives: https://www.ncbi.nlm.nih.gov/pmc/articles/PMC3328522/

"Climate Change and Mosquito-Borne Diseases: Impacts and Implications for Public Health," a report by the Climate and Health Program at the Harvard T.H. Chan School of Public Health: https://www.hsph.harvard.edu/climate-change-and-mosquito-borne-diseases/

"Water and Health: Chemical and Biological Contaminants," a resource from the World Health Organization (WHO): https://www.who.int/water_sanitation_health/publications/water-quality-guidelines-chemical-contaminants/en/

"Climate Change and Waterborne Disease Risk in the Eastern United States," a paper by Rachel G. Licker et al. published in Epidemiology: https://journals.lww.com/epidem/Abstract/2018/07000/Climate_Change_and_Waterborne_Disease_Risk_in_the.5.aspx

"Climate Change and Waterborne Disease in Developing Countries," a paper by Jamie Bartram and Suresh Kumar Rohilla published in Current Environmental Health Reports: https://link.springer.com/article/10.1007/s40572-015-0059-6

"The Impacts of Climate Change on Water and Health in the United States," a report by the Global Change Research Program: https://health2016.globalchange.gov/water-related-impacts/water-and-health

"Climate Change and Water Quality: Impacts on Drinking Water and Recreational Water," a report by the National Water Quality Monitoring Council: https://water.usgs.gov/nawqa/pubs/circulation/1331/

"Climate Displacement in America: The Tipping Point" report by the Climate Justice Program of the Center for Biodiversity: https://www.biologicaldiversity.org/climate_change/esa_climate_refugees/pdfs/Cli

mate_Displacement_in_America.pdf

"Hurricane Harvey: Facts, Damage & Impact" article by Live Science: https://www.livescience.com/60129-hurricane-harvey-facts.html

"California Wildfires: Latest Updates on Size, Containment, and Damage" article by Vox: https://www.vox.com/2018/11/19/18103110/california-wildfires-latest-updates-size-containment-damage

"Kiribati: A drowning paradise in the South Pacific" article by The Guardian: https://www.theguardian.com/world/2014/oct/21/kiribati-climate-change-pacific-island-drowning

"Desertification: Causes, effects, and solutions" article by World Wildlife Fund: https://www.worldwildlife.org/threats/desertification

"Climate change and human security" report by the United Nations: https://www.un.org/en/climatechange/climate-change-and-human-security

Chapter 4 Climate Change and Its Effect on the Global Economy

"Climate Change and Coastal Flooding Impacts" by the Environmental Protection Agency (EPA): https://www.epa.gov/sites/production/files/2017-06/documents/epa-cira-sea-level-rise-and-coastal-flooding-impacts-2017.pdf

"Climate Change and the Oceans" by the Intergovernmental Panel on Climate Change (IPCC): https://www.ipcc.ch/srocc/chapter/chapter-5-sea-level-rise-and-implications-for-low-lying-islands-coasts-and-communities/

"Coastal Erosion and Climate Change" by the Union of Concerned Scientists: https://www.ucsusa.org/resources/coastal-erosion-and-climate-change

"Ocean Acidification and Its Potential Effects on Marine Ecosystems" by the National Oceanic and Atmospheric Administration (NOAA):

gewo-intl.org/the-threat-to-earth/

https://www.noaa.gov/explainers/ocean-acidification-and-its-potential-effects-on-marine-ecosystems

"The Impacts of Extreme Heat on Infrastructure" by the American Society of Civil Engineers (ASCE): https://www.asce.org/templates/press-release-detail.aspx?id=24142

Climate change impacts on agriculture - United Nations Food and Agriculture Organization:
http://www.fao.org/3/i6307e/i6307e.pdf

Climate change and agriculture in Africa - United Nations Environment Programme:
https://www.unenvironment.org/resources/report/climate-change-and-agriculture-africa

Impact of climate change on water resources in India - International Journal of Engineering Research and General Science:
https://www.ijergs.org/papers/volume%204/Issue2/Part%204/IJERGS40204.pdf

Soil degradation and climate change - Food and Agriculture Organization of the United Nations:
http://www.fao.org/3/i1688e/i1688e.pdf

Extreme weather events and their impacts on agriculture - Environmental Defense Fund:
https://www.edf.org/climate/how-climate-change-impacting-us-agriculture

Climate Change and Tourism:
https://www.worldbank.org/en/topic/climatechange/brief/global-challenges-of-climate-change-on-tourism

Climate Change and the Tourism Industry:
https://www.unece.org/info/media/news/environment/2020/climate-change-and-

the-tourism-industry/doc.html

The Impacts of Climate Change on the Tourism Sector in Small Island Developing States: https://www.un.org/development/desa/dpad/wp-content/uploads/sites/45/publication/2018_UNDP_CLIMATE_TOURISM_REPORT.pdf

Climate Change Impacts on Coastal Tourism: A Case Study of Waikiki Beach, Honolulu, Hawaii: https://www.mdpi.com/2071-1050/12/20/8474

Increase in costs of fossil fuel production:
The Guardian: Fossil fuel subsidies are a staggering $5 tn per year, says IMF - https://www.theguardian.com/environment/2015/may/18/fossil-fuel-companies-getting-10m-a-minute-in-subsidies-says-imf

World Economic Forum: The true cost of fossil fuel subsidies - https://www.weforum.org/agenda/2020/02/the-true-cost-of-fossil-fuel-subsidies/

Higher costs of energy transportation:
Environmental Defense Fund: The Hidden Costs of Fossil Fuels - https://www.edf.org/energy/hidden-costs-fossil-fuels

Natural Resources Defense Council: The Hidden Costs of Fossil Fuels - https://www.nrdc.org/stories/hidden-costs-fossil-fuels

Investment in renewable energy sources:
International Energy Agency: Renewables 2020 - Analysis and forecast to 2025 - https://www.iea.org/reports/renewables-2020

Forbes: Renewable Energy Is Now the Cheapest Option - Even Without Subsidies - https://www.forbes.com/sites/jamesconca/2020/05/18/renewable-energy-is-now-the-cheapest-option-even-without-subsidies/

Demand for energy-efficient appliances and vehicles:
Energy.gov: Benefits of Energy Efficiency -

gewo-intl.org/the-threat-to-earth/

https://www.energy.gov/eere/buildings/benefits-energy-efficiency

CNBC: Electric cars are now less expensive to own and operate than gas vehicles, new study finds - https://www.cnbc.com/2021/04/06/electric-cars-are-now-less-expensive-to-own-than-gas-vehicles-study.html

Incentives for energy efficiency:
U.S. Department of Energy: Tax Credits, Rebates & Savings - https://www.energy.gov/savings/search?f%5B0%5D=im_field_rebate_eligibility%3A294&f%5B1%5D=im_field_rebate_sector%3A469

Energy Sage: Solar Incentives, Rebates and Tax Credits - https://www.energysage.com/solar/rebates-and-incentives/

Chapter 5 Threat of Poor Nations to Global fight against Climate Change

Climate Change and Poor Nations: How Their Situation Threatens the Global Fight against Climate Change. Retrieved from https://www.undp.org/content/undp/en/home/librarypage/poverty-reduction/climate-change-and-poor-nations.html

Bangladesh: highly vulnerable to climate change, with limited financial resources to invest in climate change mitigation and adaptation measures.
Reference: The World Bank (2021). Climate Change and Bangladesh. Retrieved from https://www.worldbank.org/en/news/feature/2019/09/03/climate-change-and-bangladesh

Ethiopia: highly vulnerable to climate change, with limited financial resources to invest in climate change mitigation and adaptation measures.
Reference: The World Bank (2021). Climate Change and Ethiopia. Retrieved from https://www.worldbank.org/en/news/feature/2019/09/03/climate-change-and-ethiopia

Haiti: highly vulnerable to climate change, with limited financial and technological resources to invest in climate change mitigation and adaptation measures.
Reference: United Nations Development Programme (UNDP) (2020). Haiti: Climate Change. Retrieved from https://www.ht.undp.org/content/haiti/en/home/operations/projects/poverty_reduction_and_environment/climate-change.html

Malawi: highly vulnerable to climate change, with limited financial resources to invest in climate change mitigation and adaptation measures.
Reference: The World Bank (2021). Climate Change and Malawi. Retrieved from https://www.worldbank.org/en/news/feature/2019/09/03/climate-change-and-malawi

Uganda: facing a severe energy crisis and limited access to renewable energy technologies.
Reference: International Renewable Energy Agency (IRENA) (2021). Uganda Energy Situation. Retrieved from https://www.irena.org/countries/Uganda

Afghanistan: facing a severe water crisis and limited access to technology that could help improve water management practices.
Reference: United Nations Development Programme (UNDP) (2019). Afghanistan: Water Scarcity. Retrieved from https://www.af.undp.org/content/afghanistan/en/home/projects/water-scarcity.html

Sierra Leone: facing a severe food security crisis and limited access to technology that could help improve crop yields and promote drought-resistant crops.
Reference: United Nations Development Programme (UNDP) (2020). Sierra Leone: Food Security. Retrieved from https://www.sl.undp.org/content/sierraleone/en/home/operations/projects/environment_and_energy/food-security.html

Yemen: facing a severe water crisis and limited access to freshwater resources.
Reference: United Nations Development Programme (UNDP) (2021). Yemen: Water Scarcity. Retrieved from

gewo-intl.org/the-threat-to-earth/

https://www.ye.undp.org/content/yemen/en/home/projects/water_scarcity.html

Somalia: facing a severe drought and food crisis, linked to climate change. Reference: United Nations Development Programme (UNDP) (2020). Somalia: Climate Change. Retrieved from https://www.so.undp.org/content/somalia/en/home/operations/projects/climate-change.html

The Paris Agreement is an international agreement adopted by the United Nations Framework Convention on Climate Change (UNFCCC) in December 2015. The official text of the agreement can be found on the UNFCCC website: https://unfccc.int/sites/default/files/english_paris_agreement.pdf

Regarding rich nations' contributions towards the fight against climate change, the UNFCCC tracks and publishes information on the progress of each country towards its commitments under the Paris Agreement. This information can be found on the UNFCCC website under the section "Nationally Determined Contributions (NDCs)": https://unfccc.int/process-and-meetings/the-paris-agreement/nationally-determined-contributions-ndcs/nationally-determined-contributions-ndcs

Chapter 6. Dooms Day Is Coming

Endangered Cities that would disappear in the next 100 years.

Intergovernmental Panel on Climate Change (IPCC). (2018). Global Warming of 1.5°C. Retrieved from https://www.ipcc.ch/sr15/

Climate Central. (2019). The World's 10 Most Vulnerable Cities to Climate Change. Retrieved from https://www.climatecentral.org/news/report-the-worlds-10-most-vulnerable-cities-to-climate-change

National Geographic. (2020). 15 Cities That Could Disappear Due to Climate Change. Retrieved from https://www.nationalgeographic.com/environment/2019/03/15-cities-disappearing-due-to-climate-change/

United Nations Environment Programme (UNEP). (2019). Adapt Now: A Global Call for Leadership on Climate Resilience. Retrieved from https://www.unep.org/resources/adapt-now-global-call-leadership-climate-resilience

Climate Central. (2019). Coastal Risk Screening Tool: Methodology Report. Retrieved from https://riskfinder.climatecentral.org/about/methodology/

Kulp, S. A., & Strauss, B. H. (2019). New elevation data triple estimates of global vulnerability to sea level rise and coastal flooding. Nature Communications, 10(1), 1-12. doi: 10.1038/s41467-019-12808-z

"New elevation data triple estimates of global vulnerability to sea level rise and coastal flooding," is available online. You can access the study through the following link: https://www.nature.com/articles/s41467-019-12808-z.

Endangered Animals and Organisms that would disappear in the next 100 years as a result of Climate Change.

Coral reefs:
National Oceanic and Atmospheric Administration. (2018). Coral reefs. https://www.noaa.gov/education/resource-collections/ocean-coasts-coral-reefs/coral-reefs
Hughes, T. P., Anderson, K. D., Connolly, S. R., Heron, S. F., Kerry, J. T., Lough, J. M., Baird, A. H., Baum, J. K., Berumen, M. L., Bridge, T. C. L., Claar, D. C., Eakin, C. M., Gilmour, J. P., Graham, N. A. J., Harrison, H., Hobbs, J.-P. A., Hoey, A. S., Hoogenboom, M., … Wilson, S. K. (2018). Spatial and temporal patterns of mass bleaching of corals in the Anthropocene. Science, 359(6371), 80–83. https://doi.org/10.1126/science.aan8048
Frieler, K., Meinshausen, M., Golly, A., Mengel, M., Lebek, K., Donner, S. D., & Hoegh-Guldberg, O. (2013). Limiting global warming to 2°C is unlikely to save most coral reefs. Nature Climate Change, 3(2), 165–170. https://doi.org/10.1038/nclimate1674

gewo-intl.org/the-threat-to-earth/

International Union for Conservation of Nature. (2021). Coral reefs. https://www.iucn.org/theme/marine-and-polar/coral-reefs

African elephants:
Wittemyer, G., Northrup, J. M., Blanc, J., Douglas-Hamilton, I., Omondi, P., & Burnham, K. P. (2014). Illegal killing for ivory drives global decline in African elephants. Proceedings of the National Academy of Sciences, 111(36), 13117–13121. https://doi.org/10.1073/pnas.1403984111
Leathem, C. S., Mitchell, D., & Lomas, M. (2019). Climate change vulnerability assessment for African elephants. Nature Communications, 10(1), 1–8. https://doi.org/10.1038/s41467-019-12481-5
Pimm, S. L., Jenkins, C. N., Abell, R., Brooks, T. M., Gittleman, J. L., Joppa, L. N., Raven, P. H., Roberts, C. M., & Sexton, J. O. (2014). The biodiversity of species and their rates of extinction, distribution, and protection. Science, 344(6187), 987–998. https://doi.org/10.1126/science.1246752

Monarch butterflies:
National Wildlife Federation. (n.d.). Monarch butterflies. https://www.nwf.org/Educational-Resources/Wildlife-Guide/Invertebrates/Monarch-Butterflies
Thogmartin, W. E., Wiederholt, R., Oberhauser, K., Drum, R. G., Diffendorfer, J. E., Altizer, S., Taylor, O. R., Pleasants, J. M., Semmens, D. J., Semmens, B. X., Erickson, R. A., Libby, K., & Lopez-Hoffman

Arctic caribou:
USGS. (2017). Arctic caribou and climate change. Retrieved from https://www.usgs.gov/centers/asc/science/arctic-caribou-and-climate-change

Koalas:
Australian Koala Foundation. (2021). Koalas under threat. Retrieved from https://www.savethekoala.com/about-koalas/koalas-under-threat
Department of Agriculture, Water and the Environment. (2021). Koala. Retrieved from https://www.environment.gov.au/biodiversity/threatened/species/koala

Golden toads:
Pounds, J. A., Fogden, M. P. L., & Campbell, J. H. (1999). Biological response to climate change on a tropical mountain. Nature, 398(6728), 611-615. doi: 10.1038/19297

Mountain gorillas:
IUCN. (2018). Gorilla beringei ssp. beringei. The IUCN Red List of Threatened Species 2018. Retrieved from https://www.iucnredlist.org/species/39999/136513435
Plumptre, A. J., Nixon, S., Critchlow, R., Vieilledent, G., Nishuli, R., Kirkby, A. E.,... & Beale, C. M. (2019). Catastrophic decline of world's largest primate: 80% loss of Grauer's gorilla (Gorilla beringei graueri) population justifies Critically Endangered status. PLoS ONE, 14(12), e0225123. doi: 10.1371/journal.pone.0225123
Gray, M., McNeilage, A., & Spelman, L. (2013). The decline of the Virunga mountain gorillas: Extent, causes and responses. Oryx, 47(04), 568-579. doi: 10.1017/S0030605312001131

Adélie penguins:
Jenouvrier, S., Caswell, H., Barbraud, C., Holland, M., Stroeve, J., & Weimerskirch, H. (2019). Demographic models and IPCC climate projections predict the decline of an emperor penguin population. Proceedings of the National Academy of Sciences, 116(49), 24526-24531. doi: 10.1073/pnas.1903589116
Younger, J. L., Clucas, G. V., Kooyman, G., Wienecke, B., Rogers, A. D., Trathan, P. N.,... & Hart, T. (2015). Too much of a good thing: sea ice extent may have forced emperor penguins into refugia during the last glacial maximum. Global Change Biology, 21(6), 2215-2226. doi: 10.1111/gcb.12876

Black Rhinoceros: "Black Rhinoceros Diceros bicornis," International Union for Conservation of Nature, accessed March 17, 2023, https://www.iucnredlist.org/species/6557/11166029.

Chinook Salmon: Jeffery J. Hard, Nathan J. Mantua, and Eric P. Salathé, "The Future of Pacific Salmon in the Face of Climate Change: Lessons from One of the Fastest Warming Basins in the World," Fisheries 37, no. 10 (2012): 432-44,

gewo-intl.org/the-threat-to-earth/

https://doi.org/10.1080/03632415.2012.719657.

Blue Whale: Miranda Jones et al., "Climate Change and Marine Mammals," in Encyclopedia of Marine Mammals, 3rd ed., ed. Bernd Würsig, J. G. M. Thewissen, and Kit M. Kovacs (Elsevier, 2018), 209-16, https://doi.org/10.1016/B978-0-12-804327-1.00129-6.

Siberian Tiger: Dale G. Miquelle et al., "Tigers and Wolves in the Russian Far East: Competitive Exclusion, Functional Redundancy, and Conservation Implications," in Wolves: Behavior, Ecology, and Conservation, ed. L. David Mech and Luigi Boitani (University of Chicago Press, 2003), 461-72.

Red Panda: Wei Wei et al., "Modeling the Effects of Climate Change on the Potential Distribution of the Endangered Red Panda (Ailurus fulgens) in the Sichuan Province of China," Biological Conservation 204 (2016): 360-68, https://doi.org/10.1016/j.biocon.2016.10.032.

The International Union for Conservation of Nature (IUCN) Red List of Threatened Species: The IUCN maintains a comprehensive list of threatened and endangered species and includes information on the potential impacts of climate change on these species. https://www.iucnredlist.org/

The World Wildlife Fund (WWF): The WWF provides information on species that are particularly vulnerable to climate change, including polar bears, marine turtles, and several species of penguins. https://www.worldwildlife.org/threats/climate-change

The United Nations Environment Programme (UNEP): The UNEP provides information on how climate change is impacting species and ecosystems around the world. https://www.unep.org/resources/report/climate-change-and-biodiversity

The National Wildlife Federation (NWF): The NWF provides information on species that are particularly vulnerable to climate change in North America, including the American pika, the northern spotted owl, and the wood turtle. https://www.nwf.org/Educational-Resources/Wildlife-Guide/Threats-to-

Wildlife/Climate-Change

The National Oceanic and Atmospheric Administration (NOAA): NOAA provides information on how climate change is impacting marine species, including coral reefs, marine mammals, and sea turtles. https://www.noaa.gov/education/resource-collections/ocean-coasts-education-resources/climate-change-and-marine-animals

Chapter 7 Addressing Climate Change: Mitigation and Adaptation Strategies

Global Carbon Project. (2021). Carbon Budget 2021. Retrieved from https://www.globalcarbonproject.org/carbonbudget/21/index.htm

International Energy Agency. (2021). CO_2 Emissions from Fuel Combustion 2021. Retrieved from https://www.iea.org/reports/co2-emissions-from-fuel-combustion-2021

https://www.bbc.com/news/world-asia-china-57483492

https://www.scmp.com/business/article/3204523/methane-takes-centre-stage-how-will-china-largest-emitter-greenhouse-gas-globally-reduce-emissions

https://www.investopedia.com/articles/investing/092915/5-countries-produce-most-carbon-dioxide-co2.asp

https://www.bbc.com/news/world-asia-57018837

Largest Producer of Solar Panels:
International Energy Agency. (2021). Global EV Outlook 2021. https://www.iea.org/reports/global-ev-outlook-2021

CNBC. (2021). China dominates solar panels but may soon face manufacturing competition. https://www.cnbc.com/2021/05/04/china-dominates-solar-panels-

gewo-intl.org/the-threat-to-earth/

but-may-soon-face-manufacturing-competition.html

Largest Installed Capacity of Wind Power:
Global Wind Energy Council. (2021). Global Wind Report 2021. https://gwec.net/global-wind-report-2021/
China Energy Portal. (2021). China's Wind Power Industry. https://chinaenergyportal.org/en/2021/06/14/chinas-wind-power-industry-2021/

Hydropower:
International Energy Agency. (2021). Key World Energy Statistics 2021. https://www.iea.org/reports/key-world-energy-statistics-2021

International Hydropower Association. (2021). Hydropower Status Report 2021. https://www.hydropower.org/status-report/2021

Renewable Energy Targets:
Xinhua News Agency. (2020). China sets 2030 targets for solar, wind power capacity. http://www.xinhuanet.com/english/2020-11/28/c_139548344.htm

Climate Action Tracker. (2021). China. https://climateactiontracker.org/countries/china/

"Russia's Energy Strategy to 2035," Ministry of Energy of the Russian Federation, 2019. Available at: https://minenergo.gov.ru/node/1144 (accessed March 28, 2023).

"Russia launches renewable energy program," Power Engineering International, 2021. Available at: https://www.powerengineeringint.com/renewables/russia-launches-renewable-energy-program/ (accessed March 28, 2023).

"Russia's Energy Efficiency and Conservation Program 2013-2020," Ministry of Energy of the Russian Federation, 2013. Available at: https://minenergo.gov.ru/node/259 (accessed March 28, 2023).

"Russia's Low-Carbon Development Strategy: Towards a Green Economy," Global Green Growth Institute, 2019. Available at: https://www.greengrowthknowledge.org/resource/russias-low-carbon-development-strategy-towards-green-economy (accessed March 28, 2023).

"Russia's climate goals need to go beyond hydrocarbons," Climate Home News, 2021. Available at: https://www.climatechangenews.com/2021/06/28/russias-climate-goals-need-go-beyond-hydrocarbons/ (accessed March 28, 2023).

"The Global Carbon Atlas," Global Carbon Project, 2021. Available at: https://www.globalcarbonatlas.org/en/CO2-emissions (accessed March 28, 2023).

"Russia: Greenhouse gas emissions", Climate Watch, 2021. Available at: https://www.climatewatchdata.org/countries/RUS (accessed March 28, 2023).

"Russia's Fourth Biennial Report to the United Nations Framework Convention on Climate Change," Russian Federal Service for Hydrometeorology and Environmental Monitoring (Roshydromet), 2021. Available at: https://unfccc.int/documents/284418 (accessed March 28, 2023).

"Russia – Climate Action Tracker," Climate Action Tracker, 2021. Available at: https://climateactiontracker.org/countries/russia/ (accessed March 28, 2023).

"Russia – Energy," International Energy Agency, 2021. Available at: https://www.iea.org/countries/russia (accessed March 28, 2023).

Adaptation Measures to Reduce Vulnerability to the Impacts of Climate Change

Intergovernmental Panel on Climate Change. (2018). Global warming of 1.5°C. https://www.ipcc.ch/sr15/

United Nations Framework Convention on Climate Change. (2019). Adaptation. https://unfccc.int/topics/adaptation-and-resilience/workstreams/adaptation

gewo-intl.org/the-threat-to-earth/

Global Environment Facility. (2021). Adaptation. https://www.thegef.org/topics/adaptation

United Nations Development Programme. (2019). Adaptation. https://www.undp.org/content/undp/en/home/sustainable-development-goals/goal-13-climate-action/targets/target-13-1/adaptation.html

United Nations Environment Programme. (2021). Adaptation. https://www.unep.org/explore-topics/climate-change/adaptation

United Nations Office for Disaster Risk Reduction. (2019). Global Assessment Report on Disaster Risk Reduction. https://www.undrr.org/publication/2019-global-assessment-report-disaster-risk-reduction

United Nations Development Programme. (2021). Sustainable Development Goals. https://www.undp.org/sustainable-development-goals

The Nature Conservancy. (2021). Green Infrastructure. https://www.nature.org/en-us/what-we-do/our-priorities/provide-food-and-water-sustainably/protect-and-restore-water-sources/green-infrastructure/

Climate Services Partnership. (2021). Climate Information Services. https://www.climateservicespartnership.org/themes/climate-information-services/

United Nations Development Programme. (2018). Community Resilience. https://www.undp.org/content/undp/en/home/librarypage/climate-and-disaster-resilience/community-resilience.html

Climate Finance Access Hub. (2021). Climate Governance. https://www.climateregistry.org/resources/climate-governance/

United Nations Human Settlements Programme. (2019). Sustainable Urban Development. https://unhabitat.org/sustainable-urban-development

World Health Organization. (2021). Climate Change and Health. https://www.who.int/news-room/fact-sheets/detail/climate-change-and-health

Caribbean Community Climate Change Centre. (2021). Climate-Sensitive Tourism. https://www.caribbeanclimate.bz/projects/climate-sensitive-tourism/

Importance of International Cooperation and Policy Action
The Net Zero Vision

United Nations Framework Convention on Climate Change (UNFCCC): https://unfccc.int/

Paris Agreement: https://unfccc.int/process-and-meetings/the-paris-agreement/the-paris-agreement

Intergovernmental Panel on Climate Change (IPCC): https://www.ipcc.ch/

United Nations Sustainable Development Goals: https://sdgs.un.org/goals/goal13

Carbon Pricing Leadership Coalition: https://www.carbonpricingleadership.org/

Global Alliance for Buildings and Construction: https://globalabc.org/about-us/

Renewable Energy Policy Network for the 21st Century: https://www.ren21.net/about-ren21/

Global Environment Facility: https://www.thegef.org/about

Climate and Clean Air Coalition: https://ccacoalition.org/en/about-us

International Solar Alliance: https://www.isolaralliance.org/about-isa/

gewo-intl.org/the-threat-to-earth/

Chapter 8 Conclusion

Intergovernmental Panel on Climate Change (IPCC). (2018). Global Warming of 1.5°C. Retrieved from https://www.ipcc.ch/sr15/
United Nations Framework Convention on Climate Change. (2015). The Paris Agreement. Retrieved from https://unfccc.int/process-and-meetings/the-paris-agreement/the-paris-agreement
United Nations Development Programme. (2021). Climate Change. Retrieved from https://www.undp.org/content/undp/en/home/sustainable-development-goals/climate-action.html
NASA Global Climate Change. (n.d.). Vital Signs of the Planet. Retrieved from https://climate.nasa.gov/
United Nations. (2019). Climate Action Summit 2019. Retrieved from https://www.un.org/en/climatechange/summit-2019.shtml

The CityGreen Blueprint – A London, Ontario Initiative

City of London. (2019). City of London Climate Emergency Action Plan. Retrieved from https://www.london.ca/city-hall/strategies-initiatives/climate-emergency-action-plan

City of London. (2019). CityGreen: London's Roadmap to Sustainability. Retrieved from https://www.london.ca/city-hall/strategies-initiatives/citygreen

City of London. (2021). Climate Emergency Work Plan 2021-2022. Retrieved from https://www.london.ca/city-hall/strategies-initiatives/climate-emergency-action-plan/climate-emergency-work-plan

City of London. (2021). Climate Action Incentive Programs. Retrieved from https://www.london.ca/city-hall/strategies-initiatives/climate-emergency-action-plan/climate-action-incentive-programs

CTV News London. (2019, September 30). London city council approves climate emergency plan. Retrieved from https://london.ctvnews.ca/london-city-council-approves-climate-emergency-plan-1.4623369